Also by Jeff Jarvis

What Would Google Do?

PUBLIC PARTS

How Sharing in the Digital Age
Improves the Way We Work and Live

Jeff Jarvis

SIMON & SCHUSTER

New York   London   Toronto   Sydney   New Delhi

 Simon & Schuster
1230 Avenue of the Americas
New York, NY 10020

First Simon & Schuster hardcover edition September 2011

Simon & Schuster and colophon are registered trademarks of Simon & Schuster, Inc.

For information about special discounts for bulk purchases, please contact
Simon & Schuster Special Sales at 1-866-506-1949 or business@simonandschuster.com.

The Simon & Schuster Speakers Bureau can bring authors to your live event.
For more information or to book an event contact the Simon & Schuster Speakers Bureau
at 1-866-248-3049 or visit our website at www.simonspeakers.com.

Manufactured in the United States of America

10 9 8 7 6 5 4 3 2 1

Library of Congress Cataloging-in-Publication data is available.

ISBN 978-1-4516-3635-2

For my wife, Tammy, my children, Jake and Julia,
my parents, Joan and Darrell, and my sister, Cindy

. . . and for Howard Stern, who inspires the title and the public life

Contents

Introduction:
The Ages of Publicness*

"Facebook," my son, Jake, told me, "was my favorite part of high school." I don't think I had a favorite part of high school. Jake's Class of 2010 was the first to use Facebook after it expanded past colleges in 2005. It extended their school life around the clock. In my youth, that would have been a formula for the infinite loop of hell. My adolescent years were dominated by hormonal high drama, a desert of irony, and social awkwardness I dare not recall or I will cringe into a small ball—all alleviated by some good friends and a few great teachers. But for Jake, Facebook allowed him to build and maintain relationships with more friends more of the time. For the Facebook Class of 2010, school became a more social experience, a good experience. As far as I'm concerned, that's a miracle.

Facebook's founder, Mark Zuckerberg, is Jake's role model—his hero, even. Jake is in college, studying computer science and entrepreneurship. While still in high school, he used his skills to write Facebook apps, turning one of them into a business he sold. Zuckerberg left college to combine those disciplines and create one of the two great corporations setting the course for our next age. I wrote about the other company in my last book, *What Would Google Do?* Just as Google built an industry around

---

*No, "publicness" is not a word in most dictionaries. I don't care. We invent words and remix definitions every day. Its once appropriate synonym, "publicity," has come to mean public relations (a phrase that itself has been freighted with marketing meaning). So please allow me to define the word: **pub·lic·ness:** 1. The act or condition of sharing information, thoughts, or actions. 2. Gathering people or gathering around people, ideas, causes, needs: "Making a public." 3. Opening a process so as to make it collaborative. 4. An ethic of openness.

search, Facebook is at the core of its own new industry, built on sharing. It is enabling and exploiting our explosive desire to connect with one another. And it is causing us to ask—as individuals and as societies—what should be private and what should be public and why. This book is not a sequel to the last; it is not *What Would Facebook Do?* It is a study of our emerging age of publicness. Here, I will examine the profound change that is overtaking us, presenting us with questions, fears—and opportunities. I will focus on the opportunities.

If my teen years were socially stunted, I'm making up for it in middle age. I have Jake to thank for much of that. He is the webmaster of my blog and my secret weapon in understanding the social age. He schooled me in the norms and values of Facebook society. He's the one who made me pay attention to Twitter, which has made writing this book at once harder—causing constant distraction with the siren call of the conversation that never ends—but also easier, as I always have researchers and editors at the ready. At this moment, I'm at my laptop, trying to catalogue the benefits of publicness. As a reflex, I turn to Twitter and ask people there what new and valued relationships they've made because they are open and public. In moments, answers flow. @john_blanton says he found his wife via chat. Lesbian comic and speaker @heathr says, "coming out brought me integrity, less fear, and more energy." An old friend, @terryheaton, says, "It helps avoid a lot of losers when dating." @flmparatta found a job. @ginatrapani created a career. @everywheretrip says he has met "people all over the world because I let people know where I am on Twitter, Facebook, and blog." @akstanwyck says that on her "last trip to nyc I made a point of meeting in person 3 folks with whom I'd bonded on twitter." In multiple tweets, @alexis_rueal says that she "found most of my friends from high school, some from college and . . . discovered that people I might not have liked 15 years ago have become wonderful friends & I cherish them all now." @sivavaid—the author of *The Googlization of Everything* and a frequent though friendly sparring partner of mine in panels and posts—responds to my question about forming valued relationships, tweeting, "how about you and me?"

Because I am public, I have made new friends and reconnected with old ones. I have received work and made money—including this book and the last. I have tested ideas, spread those ideas, and gotten credit (and blame). I'll echo @dustbury, who answered my question on Twitter by saying, "The best part about being public is that I can't BS anymore: too many are in a position to call me on it. Makes life easier." Ditto @jmheggen: "Being public led to my mantra of honesty. I am who I am all the time because, being public, lies have thin shadows." Politicians and corporations could learn from that tweet. @clindhartsen says he used Twitter and publicness to confess what he eats and weighs; "that plus self-determination has lost me 65 lbs." Not to be outdone, I have written about my malfunctioning penis—more on that later, I'm sure you'll be glad to know—and received invaluable advice in return from fellow prostate cancer patients. Being public helps me get information and make decisions. I have learned that the more we share, the more we benefit from what others share. I am a public man. My life is this open book.

Privacy advocates say I should be wary. They say I shouldn't open up so much. These privacy advocates swarm in the media every time a new online service entices us to share something about ourselves. They say we should fear the companies and technologies that use the bait of free content and services, improved social lives, personalization, and increased relevance to get us to open up. They fret about government—and they're right to, for government has the means to learn much about citizens and the power to use that knowledge against them. Privacy advocates worry for our young people, who they fear are saying too much. Bad things could happen, they warn. But then, bad things always could.

Search Google News for "privacy advocates," and in just one day you'll find no end of them quoted by media as an often-anonymous tribe of chronic worriers: "**Privacy advocates** howled." "**Privacy advocates** cry foul." "**Privacy advocates** are no doubt sharpening their arrows this morning." "Facebook angers **privacy advocates**." "E-retail law rankles **privacy advocates**." "**Privacy advocates**, civil libertarians and some social scientists are incredulous." "**Privacy advocates** will now be watching closely." "Consumers and **privacy advocates** are forever concerned about the ways they can be tracked online." They howl, cry foul, sharpen arrows, get angry, get rankled, are incredulous, are concerned, watch, and fret, our privacy advocates.

In his book *Understanding Privacy,* Daniel J. Solove compiles our reputed fears about privacy, quoting *The Naked Society*'s Vance Packard in 1964, who worried that privacy was "evaporating," and also psychologist Bruno Bettelheim, who declared in 1968 that "privacy is constantly under assault." Says Solove:

> Countless commentators have declared that privacy is "under siege" and "attack"; that it is in "peril," "distress," or "danger"; that it is "eroding," "evaporating," "dying," "shrinking," "slipping away," "diminishing," or "vanishing"; and that it is "lost" or "dead." Legions of books and articles have warned of the "destruction," "death," or "end" of privacy. As Professor Deborah Nelson has put it, "Privacy, it seems, is not simply dead. It is dying over and over again."[1]

Or is it? With all this talk of privacy, privacy, privacy, we might well end up with more protection than ever—perhaps too much. Still, I fully affirm our right to privacy, the need for it to be protected, and the need for each of us to maintain our proper control over our information, creations, and identities. I will support the armies of self-anointed privacy advocates who argue these points on our behalf—though I will try to push past the excited rhetoric, speculative fears, and vague language of privacy and examine what we mean when we talk about it. What is it that we need to keep private and why? What is the harm done when privacy is violated? What are the roots of our fears about privacy? How do we correlate one another's different expectations of privacy? Why, for example, do some Germans object to Google Street View taking pictures of their buildings while some Americans seek out the Google car to perform and have their pictures taken for all to see?

Privacy and publicness are not mutually exclusive; indeed, they depend upon each other. "Public and private are relative terms, like hot and cold or light and dark," says Paul Kennedy, host of the CBC show *Ideas.* "The one defines the other."[2] Or, as Michael Warner writes in *Pub-*

lics and Counterpublics, "most things are private in one sense and public in another."[3] A book, for example, is the public expression of private thoughts. We bring our private identities to our public acts—we decide in private where we stand on an issue, and making that public is what allows us to join with like thinkers, share our ideas, and organize action. At the same time, our public lives amid other people—hearing their ideas, arguments, and evidence—informs our private decisions. Publicness depends on privacy.

Private and public are choices we make: to reveal or not, to share or not, to join or not. Each has benefits, each hazards. We constantly seek a balance between the two—only today, technology brings new choices, risks, and opportunities. Whenever possible, we want to make these choices ourselves and not have others—companies, governments, or gossips—make them for us. As we face these decisions, I want us to be mindful not only of the risks to privacy but also of the advantages of publicness. Privacy should not be our only concern. Privacy has its advocates. So must publicness.

In this book, I will argue that if we become too obsessed with privacy, we could lose opportunities to make connections in this age of links. The link is a profound invention. Links don't just connect us to web pages, they also allow us to connect to each other, to information, to actions, and to transactions. Links help us organize into new societies and redefine our publics. When, out of fear of the unknown, we shut ourselves off from links to one another, we lose as individuals, as companies, and as institutions. When we open up, we gain new chances to learn, connect, and collaborate. Through tools ranging from TripAdvisor to Wikipedia, from Google's search to Facebook, we gain access to the wisdom of the crowd—that is, our wisdom. When we gather together, we can create new public entities—our public spheres. We must keep in mind that what's public is a public good, a necessity for an open and free society.

Government is the apparent embodiment of the public sphere. It is meant to be the agent of our public will. But we should not assume that government *is* the public. When government makes that leap, it is saying it is in the better position to make decisions about our lives than we are. Now we have tools of publicness to check government power. That is what WikiLeaks is designed to do: force secrets into the open, robbing government of unnecessary confidentiality and officials of their assumed authority to hide their information and actions. Twitter, Facebook, YouTube, and our social tools of the net—with greater and lesser success—helped the people of Iran, Tunisia, Egypt, and other countries to organize as the true publics and legitimate voices of their nations. Now governments must operate differently. Yes, they require secrecy. But outside of war, crime, and protecting the individual, there is no reason for public officials to hide what they know and do from their publics. And there is every reason for governments and their constituents to collaborate in the open, identifying and solving problems—from potholes to poverty—together. See SeeClickFix, a simple service that lets anyone in a community identify a problem: a broken park bench, say. The tool allows users to gather a critical mass of neighbors, who can demand that it be fixed. The closed, defensive local official might see this new ability as a threat, a means for voters to gang up. But wise, open pols—they do exist—are using SeeClickFix to more efficiently identify where to direct their always spare resources. Washington, D.C., and San Francisco have integrated SeeClickFix into their 311 information services, so that reports of problems are automatically passed to city agencies.[4] Self-centered residents will see the tool as a way to play gotcha with do-nothing local bureaucrats. But generous neighbors are using SeeClickFix to find problems they can fix on their own—repairing that wobbly bench without government effort and tax dollars. More and more, we will see societies form and act apart from government, crossing borders—as the Middle East's freedom fighters have, inspiring and teaching one another while the whole world watches in the open. More and more, working behind the curtain will become too costly for leaders in any society, even the tyrannical.

Companies, too, are public bodies. Whether or not they sell stock, they depend on open relationships with many constituencies: customers, employees, suppliers, partners, competitors, and communities. Just as we are coming to demand transparency from government, so will we expect more openness from companies. To date, transparency has often been a buzzword, manifesting itself as press releases with crafted messages or mea culpas when somebody screws up. That's not publicness; that's PR. The truly public company will operate in the open because publicness affords businesses a new way to work, to collaborate with customers, to reset relationships, to build trust, and to find new efficiencies—producing better products, making fewer mistakes, spending less on marketing, building better brands together. Today, the more a company opens its process to customers, the more the people formerly known as consumers can move up the design, sales, and service chains to say what they want in a product before it is made.[5] Even Coach, the high-end fashion-accessory company, opened its doors, inviting bloggers to design bags, which let the company tap into a new (and less expensive) source of talent while getting free marketing from the designer-bloggers and also defanging critical bloggers.[6]

For companies, transparency can spark a virtuous cycle: Publicness demonstrates respect, which earns trust, which creates opportunities for collaboration, which brings efficiency, reduces risk, increases value, and enhances brands. Publicness is good business.

The rule can apply even in show biz. In online discussions, Tim Kring, creator of the TV series *Heroes,* saw fans criticize the direction of the show's plot. He publicly admitted missteps with the characters they loved. He fixed the problems. He gave the fans credit and thus respect. In the snaily production cycle of network television, it's difficult to collaborate with the audience in time to change the course of an entire series. Online, it's easier. Each week, I appear on a podcast called *This Week in Google* started by former TV broadcaster Leo Laporte, who set up his own network of shows on the net, each streamed live alongside a chat room where viewers talk among themselves. When Leo, our fellow panelist Gina Trapani, and I don't know something, we call to the chat room and can be guaranteed to get an answer in seconds. This arrangement makes not only for better shows but for rich and loyal relationships with fans, who tell us what they want in a show and thank us when we give it to them. Leo's two-way trust with his public is astonishing. That rubs off on his sponsors, which Leo will tout only if he trusts them. Virtuous cycle.

All these opportunities are made possible—and amplified and accelerated—by technology, by the internet, which is our new public place. My colleagues in media have tended to see the net, godlike, in their own image, as a medium. But the net is not just a medium for content. It is a means of connection. Doc Searls, a coauthor of *The Cluetrain Manifesto,* the seminal work on internet culture, says we should think of the internet as a place.[7] It is our town square, where we connect with one another. French Foreign Minister Bernard Kouchner wrote in the International Herald Tribune that the internet is "an international space."[8] The chief technology officer of the U.S. Veterans Administration calls the internet "the eighth continent."[9] A reader of my blog doesn't like the idea of the internet as a spot on the globe. He argues that it is instead a new and parallel universe; it is *that* different. I'm coming around to that point of view, that the internet is a new layer on the world, perhaps a new society, or a path to a different and more public future.

Young people live in that public future—often to the horror of their elders—because they see the rewards that come from being open. They interact in public. That is how they share and connect with one another, how they build their reputations, careers, and brands. They are savvy about the benefits and risks and, as I will show later, are learning to act accordingly, protecting their privacy with more skill and intelligence than we assume. We should learn from them, for the future is theirs.

Publicness, though, is not the sole domain of youth. By the hundreds of millions, across every age and most every connected culture, we are sharing. Unconnected cultures are coming online, too, as today's 2 billion internet users will soon be joined by 3 billion more who will get on the net via ever-less-expensive and ever-smarter phones. Do not think that the United States is the heart of the net. Brazil has long been an unsung hotbed of interactivity, early to adopt blogging, photo sharing, and friend services. China Mobile has 600 million customers (which happens to be almost as many as Facebook has—and they're not the same people). Poor farmers, fishermen, and merchants in Africa and India are using connected technology to improve their markets. About 70 percent of Facebook users come from outside the U.S.[10]

All around the world, we are already living increasingly public lives, sharing our thoughts, photos, videos, locations, purchases, and recommendations on Facebook, Twitter, Flickr, YouTube, Foursquare, and platforms offered by other companies in the sharing industry. These people aren't sharing all this because they're reckless exhibitionists, mass narcissists, senseless drunks (well, not usually), or insane. They are doing it for a reason: They realize rewards from being open and making the connections technology now affords.

Technology may bring these opportunities. But technology also breeds fear. Again and again in history, technology has caused change and that change has sparked worries that privacy is being threatened or that publicness is being thrust upon us. The invention of the printing press did that five centuries ago, as did the invention of the camera a century ago and countless other technologies since. At the start of this millennium, it is the internet that feeds these fears. So this is a story not just of privacy and publicness but also of technology and change, fear and opportunity, and the shape of a new era. Publicness is not merely an online fad, a few cool tools, a new business method, a flash of political rhetoric, a fancy of youth. No, publicness is at the heart of a reordering of society and the economy that I believe will prove to be as profound as the one brought on by Johannes Gutenberg and his press.

Technology is forcing us to question centuries-old assumptions about the roles of the individual and society: our rights, privileges, powers, responsibilities, concerns, and prospects. That describes nothing so much as the process of modernization. In ancient times, Richard Sennett says in *The Fall of Public Man,* "public experience was connected to the formation of social order"—that is, the end of anarchy. In recent centuries, being public "came to be connected with the formation of personality"— that is, with our individuality and freedom.[11] Ancient and authoritarian regimes told people what they must think and do; modern societies enable and ennoble citizens to do what they want to do, alone and together. Publicness is a progression to greater freedom. We use that freedom to express ourselves as individuals and also to find like minds and collect into new societies.

Society infrequently but inevitably splinters and splits and then reshapes into new forms. Think of us as atoms in molecules. Centuries ago, our molecules were villages and tribes; location defined us, and often religion guided us. In Europe, Gutenberg empowered Martin Luther to smash society apart into atoms, until those elements re-formed into new societies defined by new religions and shifting political boundaries. With the Industrial Revolution—of which Gutenberg himself was a first faint but volatile spark—the atoms flew apart again and re-formed once more, now in cities, trades, economies, and nations. We atomize. We re-form into new molecules. We don't evolve so much as we blow up in wrenching bursts of violence, breaking strong, old bonds, making us feel disconnected until we can connect again.[12] I am not proposing to debate whether we are meant to be alone or together, whether our natural state is independent or social, private or public. We are meant to be both; we just change the formula, given chance and necessity. We like to think that we finally find the right balance and discover our natural and permanent state. Then technologies come along and ruin our dear, old assumptions and order.

Today, the internet atomizes us once again. The net is everyone's printing press. I don't mean it is media; I just said it's not. I mean it is our tool of disruption, a catalyst that breaks old bonds and sets us loose to explore our natures anew. This transformation takes trivial form: All of us no longer watch the same, shared news with the same, one-size-fits-all viewpoint. RIP, Uncle Walter. This transformation also takes momentous form: revolutions, dead industries, economic upheaval. We atomize. We re-form. We want to be apart—too far apart, some fear. In the book *Bowling Alone,* Robert Putnam worries that we are becoming disconnected from family, friends, neighbors, and society. But then we want to be together. That book inspired entrepreneur Scott Heiferman to found Meetup, a platform that lets groups organize gatherings in person around whatever interests they share, from dogs to dance, sci-fi to science. Atomize. Re-form. We can now find the publics we wish to join based not on the gross labels, generalizations, and borders drawn about us by others—red vs. blue, black vs. white, nation vs. nation—but instead on our ideas, interests, and needs: cancer survivors, libertarians, revolutionaries, Deadheads, vegetarians, single moms, geeks, hunters, birders, and privacy advocates.

Publicness is an emblem of epochal change. It is profoundly disruptive. Publicness threatens institutions whose power is invested in the control of information and audiences. That is why we hear incumbents protest this change and warn of its dangers. Publicness is a sign of our empowerment at their expense. Dictators and politicians, media moguls and marketers try to tell us what to think and say. But now, in a truly public society, they must listen to what we say, whether we're using Twitter to complain about a product or Facebook to organize a protest. If they are to prosper, these institutions must learn to deal with us at eye-level, with respect for us as individuals and for the power we can now wield as groups—as publics. If they do not, they may be replaced by entrepreneurs or insurgents, good or bad.

The progression toward a more public society is apparent and inevitable. Resistance is futile. But the form our new society will take is by no means predestined. We are at a critical moment with many choices. We who hold the tools of publicness hold keys to the future. We must decide how to use them. Rather than baying at the moon and cursing the tide, we would be wise to find opportunity, to decide the kind of future we want to build. How can we change governments, organize politics, win votes, and acquire power with our new tools? How can we find power apart from government to help us prevent regimes from using those same tools to spy on and subjugate us? How can a company improve and profit by opening up its information and its processes to transform relationships, collaborate, and profit? At the same time, how can we ensure a company will maintain our trust by protecting our privacy? How can we empower our children to take advantage of the new and incredible options they have to create, share, and connect while also teaching them how to protect themselves from bad actors and unintended futures? These are questions I will grapple with in this book.

Now is the moment and we are the people to give shape to our next society. In our roles as individuals, parents, employees, employers, citizens, officials, and neighbors, each of us is deciding how private to be (safe, protective, closed, sometimes solitary, often anonymous) and how public (open, collaborative, collective, and vulnerable).

Like many, I witnessed the drama of Egypt's revolution play out on Twitter. Silly little Twitter. It was supposed to be made for nothing more than sharing the narcissistic trivia of our lives as we each answered the simple question: What are you doing now? As if the world should care, right? During the revolt, I tweeted how jarring the contrast could be between the everyday updates of people I knew—meals, dates, complaints, cats—flowing next to the messages of courage, fear, exhilaration, and determination I saw from the people of Cairo's Tahrir Square, strangers I was coming to know and respect by the minute. @ghonim—former Google executive Wael Ghonim, who was credited with helping spark the revolution with a Facebook page[13] and was jailed as a result—used the tools to deliver news, inspiration, and support. "Pray for #Egypt," he tweeted. "Very worried as it seems that government is planning a war crime tomorrow against people. We are all ready to die." On the seventeenth day of the eighteen-day revolution, when the people of the square believed the dictator Hosni Mubarak would leave, Ghonim tweeted too soon, "Revolution 2.0: Mission Accomplished."[14] Mubarak did not leave that night. But the next day, he did. "Welcome back, Egypt," Ghonim tweeted. "They lied at us. Told us Egypt died 30 years ago, but millions of Egyptians decided to search and they found their country in 18 days." Ghonim went on CNN to thank Facebook. "This revolution started online. This revolution started on Facebook," he said. "This revolution started in June 2010 when hundreds of thousands of Egyptians started collaborating on content. We would post a video on Facebook that would be shared by 60,000 people on their walls within a few hours. I always said that if you want to liberate a society just give them the internet."[15]

At every minute in this historic story, it was evident how precarious the next minute would be, the one after that only more so. Mubarak might or might not leave sooner or later. The army could shift this way or that. Thugs with rocks could return to the square. Authorities could use the internet to spread misinformation and find and arrest the protestors. Out of habit, I watched the news progress on TV. Most of the time, even Al Jazeera English had only a telephoto shot of the square from a safe distance, as commentators could do little more than repeat themselves. TV could hear few voices from the square. But Twitter delivered those voices, in the midst of their revolution. Blogger Doc Searls said that in Egypt, Twitter did to cable news what cable news had done to newspapers: It made an old medium less immediate.[16]

On Twitter, a virtuoso of the form, @acarvin—National Public Radio social-media strategist Andy Carvin—spent hours and days curating the best he could find from the people on the ground in Egypt. At the height of the revolution, he tweeted and retweeted 1,300 times in 24 hours. He verified who was there through his trusted sources. He passed on news, debunked rumors, and asked the people who were there what was really happening. He quoted—retweeted, that is—people such as @sandmonkey, a brave blogger who six years before had begun to share his ideas and experiences in Egypt at not inconsiderable risk. After the protestors' victory, Sandmonkey blogged, "Tonight will be the first night where I go to bed and don't have to worry about state security hunting me down, or about government goons sent to kidnap me; or about government sponsored hackers attacking my website. Tonight, for the first time ever, I feel free . . . and it is awesome!"[17] Sandmonkey replaced his anonymous cartoon avatar on Twitter with a picture of the real him and published his name, Mahmoud Salem. He was free at last to become a public man.

As you read this, months or years after the revolution began, it is, of course, too soon to know how this story will turn out. It has no script and not even a dramatis personae of leading actors. What kind of society Egypt can build and maintain post-Mubarak teeters on so many risks, needs, and warring interests, but also on so many new opportunities. Just as Egypt's society of the future could go many ways, so could ours and other new societies yet to emerge.

The new age has its doubters. Author Malcolm Gladwell plays the cur-mudgeon. "Surely," he says, "the least interesting fact about them is that some of the protestors may (or may not) have at one point or another employed some of the tools of the new media to communicate with one another. Please. People protested and brought down governments before Facebook was invented." I have no doubt that the tools of publicness played a role in helping a true Egypt of the people rise from silence to be heard at last. These tools helped them share their information, their frustration, and their dreams. That is why Mubarak shut down the inter-net and mobile phones, because these technologies posed a threat (and the fact that any one man could do that should worry us all). But even Mubarak had to turn the internet back on, because the net is that vital to life now. Yet Gladwell is right about this much: The tools are just tools. The revolution is the people's. As a blogger reminded us on Al Jazeera English, Twitter did not fight Mubarak's police; Egyptians did. Facebook will not create a new society, but this new society has used Facebook to begin to shape itself.

"What kind of world would make the values of both publicness and privacy equally accessible to all?" asks Michael Warner.[18] That is our chal-lenge: to find a new balance between our roles as free individuals and as members of a public who join together to build better, more open, more generous, and more accountable companies, markets, communities, gov-ernments, schools, relationships, and lives. There is a need for privacy, its cautions, and its advocates, to be sure. But publicness also needs its advocates. This book is one of them.

The Prophet of Publicness:
Mark Zuckerberg

I brought son Jake with me to interview Mark Zuckerberg in Facebook's Palo Alto headquarters. As we entered his glass-walled conference room, Zuckerberg rushed in nervously and said he needed to erase the whiteboard. I cursed myself for not having read it first. Zuckerberg's secrets were safe from me. It's too easy to find such conspicuous irony at Facebook: The company wants us all to open up with each other, but it is secretive. Its founder wants us all to be social, but Zuckerberg himself—as someone who knows him well once joked to me—is antisocial. Zuckerberg's mystique is not that he's so public but that he is so mysterious. He is an enigma wrapped in a nerd becoming a mogul.

As he erased, I looked out on a mountain range of oversized Dell monitors. One screen stood alone, facing us with a giant, red "45" and a countdown clock ticking below. What's happening in forty-five days? I asked. "That's when we launch a lot of stuff," Zuckerberg said, revealing nothing. He has a talent for saying as little as possible. That's why I have enjoyed watching him in high-powered panels amid mighty moguls at the World Economic Forum in Davos, Switzerland, or at Rupert Murdoch's corporate retreat in Monterey, California.[1] Zuckerberg always said what he thought, not a word more. He wouldn't blather on, spouting PR flowers to fill time and suck up attention. He didn't try to charm his questioners or his powerful audiences. He was unfailingly direct, politely blunt. He had not been ruined by media training, not yet. I also think he was a bit scared in the spotlight or at least didn't enjoy it. He said what he had to say and was impatient to move on. That's what gave him the look and reputation of the geek caught in the headlights, that and his laser-to-the-eye stare. Some have made amateur diagnoses of Asperger's syndrome.[2] Even if I were qualified to do so, I wouldn't. For Zuckerberg has changed since then. One-on-one, he is serious but subtly charming. He remembers to smile. In large groups, he is getting more comfortable, at ease interviewing even Barack Obama. Still, I couldn't get anything more out of him about that clock.

When I posed the riddle of the clock on Twitter, better mathematicians than I calculated that forty-five days hence was the date of the premiere of *The Social Network,* the 2010 film about Zuckerberg and his creation. When I saw it, I didn't much like the movie, which I thought was an attack on Zuckerberg, geeks, entrepreneurs, the internet, and change: the revenge on the revenge of the nerds. The scriptwriter, Aaron Sorkin, admitted that he didn't know anything about Facebook or much care about the facts. "I don't want my fidelity to be to the truth," he told New York Magazine. "I want it to be to storytelling." The film was fiction, wishful fiction that wanted to believe this internet thing is not a revolution but merely the creation of a few odd machine-men, the boys we didn't like in college. *The Social Network* was, in the words of New York's Mark Harris, "a well-aimed spitball thrown at new media by old media."[3]

Jake disagreed. He left the movie inspired to start a company. In random polling, I found generational differences in *The Social Network* as a Rorschach test: People my age were more likely than I to love the movie and dislike Zuckerberg; people Jake's age were more prone to like the movie and see him as a victor. I split the difference. I disliked the movie, but I do admire Zuckerberg. I believe he has a vision for a connected world and the skill and determination to realize it. Motive matters. If, as the movie paints him, he acts only out of his own cynical goals—getting attention, getting rich, getting laid—then manipulating us to reveal ourselves smells of exploitation. But if instead he has a higher aim—to help us share and connect and to make the world more open—then it's easier to respect him, as Jake and I do.

"I'm in the first generation of people who really grew up with the internet," Zuckerberg tells me. "Google came out when I was in middle school. Then there was Amazon and Wikipedia and iTunes and Napster. Each year, there were new ways to access information. Now you can look up anything you want. Now you can get cool reference material. Now you can download any song you want. Now you can get directions to anything. The world kept on getting better and better."

In his words, we hear the inherent optimism that fuels the likes of him: that with the right tools and power in the right hands, the world will keep getting better. "On balance, making the world more open is good," Zuckerberg says. "Our mission is to make the world more open and connected." The optimist has to believe in his fellow man, in empowering him more than protecting against him. Zuckerberg believes he is helping us share, which will make the world more public and lead to greater transparency and trust, accountability and integrity. That, he says, is why he started Facebook—not, as *The Social Network* would have us believe, to get a girlfriend (he already had one, the same woman he still dates, Priscilla Chan) and not, as others say, because he is trying to force us into the public. He contends he is creating the tools that help people do what they naturally want to do but couldn't do before. In his view, he's not changing human nature. He's enabling it.

"In the world before the internet and things like Facebook," he says, "there was a huge amount of privacy through obscurity." We didn't have the choice and power we have now: "We had this culture where you were either a producer or a consumer. . . . It was a very bifurcated society— kind of unnatural." That is, the tools of publicness, including the media, had been in the hands of the few; now they are in the hands of all. "So now the question isn't, 'Are you completely private?' It's, 'Which things do you want to share and which things do you not?'"

Zuckerberg says he wants to give his users control over what's public and private. If that's true, why does he keep getting in trouble regarding privacy? A few reasons: Some people complain that there aren't enough privacy controls on Facebook. So the company adds controls. But then the complaint becomes that the controls are too complicated. When they are so complicated, users tend not to bother to adjust them and instead rely on Facebook's default settings. Facebook has surprised its users with changes to those defaults, making them ever-more public—thus making its users more public, often whether they know it or not. And Facebook hasn't been good at communicating with its users. When I tell Zuckerberg my thesis for this book about the benefits of sharing, he admits, "I hope you have better luck than we've had making that argument. I think we're good at building products that hit the desire people have, not necessarily at expressing in English what the desire is." Some would call that understatement.

When Facebook creates new products, it has a tendency to release first, ask questions later, dealing with users' displeasure and revamping or retreating after the fact. It's not so much a beta method (release a bit at a time) as a deep-end method (throw all the users into the pool and look for fingers). That's what happened in the service's first major kerfuffle in 2006 with the introduction of the so-called News Feed, which compiled friends' updates and fed them onto other friends' pages. That freaked and creeped some users. Even though each individual update was already visible to members of a network as they clicked from one friend's page to another, seeing all these tidbits automatically aggregated and redisplayed surprised and worried users. Now they saw their new relationship statuses or conversations pushed out to their friends as if they were news. Many objected, starting protest groups and using Facebook's own tools against itself. Zuckerberg enjoys the irony that the protest against News Feed spread through News Feed. "So this person made a group," he says, "and the group grew because these people could see that a lot of people were joining the group through News Feed." In *The Facebook Effect*, David Kirkpatrick says that one protest group—among five hundred—grew to 700,000 members in a day. Zuckerberg regrouped. He apologized to users. "We really messed this one up," he blogged. "We did a bad job of explaining what the new features were and an even worse job of giving you control of them. . . . We didn't build in the proper privacy controls right away. This was a big mistake on our part, and I'm sorry for it."[4] He made adjustments, allowing users to control what did and did not go into the feed. But overall he held firm, stubbornly believing that users would like News Feed. He turned out to be right. News Feed became the addictive drug in Facebook's bloodstream, increasing its traffic by billions of page views shortly after its launch.

In the next dustup, Zuckerberg did surrender. His ad service, Beacon, shared users' purchases with friends. What better marketing could a product get than a friend's recommendation? But Beacon was carelessly designed. In one case, a wife learned about a ring her husband had bought her before he gave it to her. Zuckerberg thought people had demonstrated a desire to share things they did outside Facebook, such as telling friends what movies they'd seen. It was a mistake to make the same assumption about more sensitive and personal information, including something as private as purchases in Beacon, without users' knowledge and consent. Facebook's worst move was to surprise its people. After a storm and some adjustments, Zuckerberg gave up and killed Beacon. I tried to suggest to him that there was still the germ of a vision there about the future (and the death) of advertising, as friends' recommendations become the most authentic form of marketing. I'll tell you later about a service called Blippy that lets its users share purchases directly from their credit cards with friends or the world. Zuckerberg said that Blippy was right to be explicit but Beacon was just wrong because it unsettled users.

In its next and most resounding flap—at least within the small world of social-media gurus—Facebook quietly changed its privacy defaults one too many times. Blogger Matt McKeon drew a much-linked-to web graphic that displayed the progression of Facebook's default settings over the years.[5] In 2005, he showed, the default had your photos and Wall posts—things you say on your Facebook home page—visible only to friends and your network (at the time, that meant your college); by 2010, the default made this content open to all of Facebook and the entire web. In 2005, your list of friends was open to your other friends and the network; in 2009, that expanded to all of Facebook and then to the web. So Zuckerberg was accused of baring users' lives and souls for them. This time Facebook responded by creating a simplified, single page for privacy settings.

I believe Facebook wasn't so much dastardly as befuddled about the definition of public. It conflated the idea of creating *our* public (our closed circle of friends on Facebook) with speaking to *the* public (that is, the web and the world). Facebook made the tools that let each of us create and manage our own groups of friends, our own societies—*our* private publics. Other tools—Blogger, Twitter, Flickr, YouTube—let us publish and broadcast to anyone and everyone—*the* public. When Facebook confused the two sorts of publics—taking updates we thought were to be seen only by *our* public and opening them up to *the* public— it confused its users, making some fear that what they'd said to a few friends could now be overheard on the world's PA system. At this early moment in the birth of our newly public society, when most of us are just learning how to talk to the world, confusion is frightening. "Digital vertigo" is how author Andrew Keen describes the feeling in a tweet (and the title of his book).[6]

Zuckerberg, his coworkers, and their generation can't seem to grok,[7] as they would say, the depth of these fears. They grew up with the net and its culture of openness and have figured out how to manage it. "I just think everyone thinks through this stuff," says Zuckerberg. He and his cohort are sometimes called digital natives because they are supposed to understand the land and the language of this new world better than immigrants my age. That phrase, "digital natives"—popularized by traditional publishing mogul Rupert Murdoch in a 2005 speech to newspaper editors[8]—has fallen into disfavor, as it implies that young people are born with the knowledge and skills to protect themselves online. But they do need to learn those things. And I'm not sure adults are necessarily their best teachers. Given the space, trust, and respect, young people teach one another as they develop their new society's norms.

Facebook no longer serves just college students as it did at its founding in 2004, when Zuckerberg remembers drawing what he calls the social graph—the map of our connections and how information travels through it—on a whiteboard in his Harvard dorm room. "Now," he says, "the audience is everyone." So Facebook has to anticipate the desires of everyone. His goal, then, is that "right out of the box, Facebook is set up so you're sharing with all the right people and you don't actually have to go in and tweak a lot of stuff. . . . It's not 500 million people's jobs to think about all the intricacies of the system." In short: Trust the defaults. I'd think it would be easier to pull off one-size-fits-all defaults in a structured college society—where the class of 2005 doesn't have much to do with the class of 2015—than in real life, where relationships are, to use Facebook's own word, "complicated." But Zuckerberg contends that people's uses of Facebook don't vary much by age or nation. "The reason why Facebook is so universal is because everyone, for the most part, has friends and family and they want to stay connected with those people." He says Facebook tries to take some information that's not very sensitive—your name, your list of friends—and make that open by default, leaving the rest to be seen only by friends, while also trying not to bother friends with too much information about others. In its quest to maximize privacy control and minimize privacy controversies, though, I wonder whether Facebook is closing up too much. When some people ask to befriend me and I try to find out who they are, they restrict me from seeing that information. Some friends they are. In this discussion, we can lose sight of the idea that Facebook is fundamentally a place to share.[9]

In their efforts to motivate us to share, Facebook, Google, and other net services have a common, technohuman goal: to intuit our intent. They want to gather signals about us so they can tailor their content, services, and advertising to us. These services compete to find more ways to get us to generate signals—our locations, needs, tastes, relationships, histories—so they can recommend, say, the perfect restaurant for each of us, knowing where we are right now and what we like and who our friends are and what they like (and making money by giving us a well-targeted coupon for the establishment). These services come into conflict with privacy advocates because capturing and analyzing our signals to predict our desires can look to some like spying or mind reading. "How did you know I was going to France?" the skittish user wonders of Google. Likely because you searched for Paris, sir.

Zuckerberg believes that by giving you back information about your own life—your friends and what you and they like and do—you'll get "a much clearer sense of what's going on around you, allowing you to learn things you couldn't otherwise—and just be better at being human." The hubris is impressive: making better humans. Google merely wants to organize our information. Zuckerberg sees Facebook as a next step in the net's evolutionary scale toward humanity. "They crawl the web," he says. "But there's nothing you can crawl to get information about people. It's all in our minds. So in order to have that service, you need to build the tools that let people share." He identifies what he contends is another difference between Facebook and its predecessors: "All the information that's about you on Facebook, you chose to put there.[10] The last wave of sites before that do not work that way." Ad networks collect information about you from your behavior—in most cases anonymously—so they can target ads, but the process is opaque. "On Facebook, you get an ad about Green Day because you *said* you like Green Day. . . . I think these models where people have more control over stuff are going to be so much more powerful and expressive." As he talks, I come to think of Google as the third-person web; it's about others—*them*. Facebook endeavors to be the first-person web; it's about *me* and *us*.

Zuckerberg has created an asset worth billions of dollars. Wall Street laughed—but the Valley didn't—when Microsoft invested in Facebook in 2007 at a reported $15 billion valuation. By 2011, pundits pegged the value at $20 billion, $50 billion, even $100 billion. I believe he's building something even bigger, with data as a new currency: We trade information about ourselves for information about what we want. Our reward is relevance. Zuckerberg disagrees with me, saying my thinking is not "the right frame. . . . I really think it's more interaction-for-interaction than data-for-data." Keeping users' motives in mind is critical. Early location services—Google Latitude and Loopt—asked you to broadcast your location to the world without much reason to do so (and with some good reasons not to). Later services such as Foursquare and Facebook Places let you alert friends where you are so you can meet up. You interact with Facebook, telling it what you are up to, and in return you interact with friends. Interactions-for-interactions.

Zuckerberg contends that Facebook is not just a technology company but also a sociology company. I find that revealing. He's not so much an engineer—he majored in both computer science and psychology—as he is a social engineer, building systems for humans, helping us do what we want to do . . . and what he wants us to do. Take Facebook friend lists. No one wants to sit down and make a list of friends. People say they want to—in Zuckerberg's words—"subgroup their friends." But in practice, who wants to bother? I have tried to subgroup contacts in my address book—fellow geeks here, journalism colleagues there, family here—but it's tedious and I quickly give up. When you friend someone on Facebook and they friend you back, you end up with a list of your friends as a by-product. The reason you do it, Zuckerberg says, is because "it's like a cool handshake. And then it's the sum of 10 billion of those." Once unlocked from their privacy, these small acts of publicness add up. "Some people just assume that being private is good," Zuckerberg says, "whereas we've always come out saying no, no, people want to share some things and keep some things private and that'll always be true. And as time goes on and more people find that it's valuable to share things, they might share more things." That is how he designs his systems, to make it fun and beneficial to share more and more.

Zuckerberg has his own, social version of Moore's law[11]—I call it Zuck's law, though he doesn't. It decrees: This year, people will share twice as much information as they did last year, and next year, they will share twice as much again.[12] Facebook will expand to more users—from 750 million today to a billion soon?—and users will expand their sharing. Meanwhile, one Facebook investor, Yuri Milner, tells me that advances in artificial intelligence will get better and better at understanding and making use of all the service's data. It has only just begun. "The default in society today still is, OK, I should not share it. The by-far default today is that everything's anonymous," Zuckerberg laments. "In the future, things should be tied to your identity, and they'll be more valuable that way." There is the master plan.

Public Choices

Private Germans

The Germans are a deeply private people. That's not just because of their political history in the last century. I think it comes from something deeper in their soul. My German grandfather-in-law jumped ship in the United States in 1923 and brought with him an abiding sense of privacy. "You mustn't tell people that," our Opa used to say, wagging his crooked, disapproving, and protective finger. "No one needs to know that." Germans are hardly alone in grappling with issues surrounding privacy. But the debate in Germany is more strident than anywhere else I can find, making it an ideal laboratory in which to study our longing for privacy amid the inexorable shift to publicness brought on by the social net.

Google Maps' Street View has caused quite the fuss in the country. Since Google's camera-equipped cars took to German streets, there has been an escalating chorus of complaints from government regulators and media. Politicians demanded that the faces of people photographed on those streets and even the images of license plates and building numbers—all captured in public—be obscured, which Google did. In Hamburg, Google agreed to notify neighborhoods before the Street View car came to town (better hide *Oma und die Kinder*! Google's in town!). Germany's federal minister of food, agriculture, and consumer protection, Ilse Aigner, told the German newsmagazine Focus in 2010 that Street View was "a comprehensive photo offensive" that "is nothing less than a million-fold violation of the private sphere."[1] She wanted Google to obtain the consent of each citizen before posting photos of the fronts of their homes. Her office provided an online form for citizens to demand that Street View photos of their homes be obscured. I saw interest in Germany for a U.S. researcher's software that automatically erases only people from Street View images—leaving a ghostly, neutron-bomb landscape where the people disappear but the buildings remain. (It wasn't perfect; one could still see the occasional dog and leash with no owner.)[2] Politicians in the city of Leverkusen proposed charging media companies such as Google €150 for every kilometer Street View pictured.[3] It didn't help Google's cause when the company admitted that its cars had not just been taking pictures but were capturing data from the Wi-Fi networks they passed in Germany and elsewhere. The capture of that data was an inexcusable screw-up and a PR calamity, but it was hardly a conspiracy aimed at killing privacy. There was no conceivable commercial use for the bits of communication and addresses passing over residents' open Wi-Fi connections on a random moment on a random day on a random German street. Nonetheless, hostility ensued.

In 2010, the Google Street View car was vandalized in Germany. In Austria, a seventy-year-old man threatened the car with a garden pick.[4] A comedy show on the German TV network ZDF aired a spoof about a new service: "Google Home View." A Google Man in a Google hat with a camera in hand tells unsuspecting Germans at their front doors that Google is going to take pictures *inside* their homes—and some acquiesce. "You're doing this all over Germany?" asks the resident. "Every house all over Germany, every room," says Google Man. "Everything is going to be photographed, and everything is going to be on the internet." Sounds plausible enough. Because of data protection laws requiring the pixelization—the digital obscuring—of faces, Google Man gives one home's residents black bars to hold in front of their eyes—"pixel boards," he calls them. At another house, a woman turns the tables on Google Man and takes a picture of his car. Google Man holds the pixel board in front of his eyes and threatens, "Well, if you're not going to play along, we'll discontinue your Google!" Judging from the laughter in the studio, the audience got the joke even if the subjects—and regulators—didn't.[5]

If only the privacy fight over Google were always so apparent in its irony. In 2007, Germany's government debated a law that would have required Google to retain the verified names and addresses of Google users and their data to aid in criminal investigations. So on the one hand, German politicians said Google was violating citizens' privacy by gathering data. On the other hand, they were demanding that Google hold on to citizens' private communications should government wish to use that information against them. Google, appropriately horrified, threatened to shut down its email service rather than enable government spying on citizens in a country that had all too much experience with the practice.[6] A common explanation of Germans' passionate advocacy of privacy is, of course, that the secret police of the Nazis and the East German Stasi spied on citizens in their private lives. But in this case, Google proved the better protector of privacy than the modern German government.

By the time Street View opened in Germany in late 2010, 244,000 people had submitted the necessary forms to demand that Google pixelate their homes and even offices—that is, 3 percent of the 8.5 million homes in the twenty cities where the service launched.[7] As is their habit, the Germans invented a word for this: *Verpixelungsrecht,* the right to be pixelated. Mind you, not every German favored obscurity. On Twitter, some Germans mocked the furor, renaming their land Blurmany. Writer Jens Best turned it all into a game, starting a site called *Finde das Pixel* (Find the Pixel), where he challenged users to search Street View for addresses that had been blurred. He further suggested that they take their own pictures of the obscured buildings and link to them from Google Street View.[8]

I, too, had ridiculed the idea of the *Verpixelungsrecht.* But when the adulterated Street View debuted, I was less amused than appalled. I took to my blog, crying, "Germany, what have you done? You have digitally desecrated your cities."[9] Navigating a lovely German landscape online, one is suddenly assaulted with a fog of pixels obscuring the public view. At a forum on privacy held by the Green party in Berlin, where I spoke, a member of the audience asked whether future historians would blame the current generation for leaving German cities in digital ruins as bombs did the real landscape in World War II. The topic is nothing if not emotional.

Germans, I said at the conference, should indeed be mindful of their history when they try to censor Street View, for in limiting Google's ability to photograph public streets, they set a dangerous precedent: If Google can be pressured not to take a picture of a public place, what is to stop some powerful miscreant caught in a bad act from making the same demand of a journalist or a citizen? Ah, my German debaters have said, but there's a difference between a public person and private person, a public act and an act committed in public with an expectation of privacy. That is a dangerous distinction if it is made case-by-case. What if I walked out on the street and littered, and you wanted to take a picture of me to illustrate ugly Americans schmutzing your fair city? What if I claimed an expectation of privacy? Would you agree? What if, instead, it's the mayor sneaking into an opium den you catch on camera? Or a police officer beating an innocent citizen? Or a parent harming a child? Would you agree now that they may have an expectation of privacy in public (which, indeed, is what may embolden them to misbehave)? Limiting what's public may grant the tyrant a curtain to hide behind. What's public is public and should remain so. A German court has agreed. In 2010, a woman sued Google, arguing that Street View might violate her privacy. The following year, in what was called a landmark decision, Berlin's state Supreme Court ruled that Street View is legal because its pictures were taken from the street.[10]

On a visit to Berlin, I sat in the offices of Bild, the largest newspaper in Europe. It is the populist paper—a tabloid, we'd call it in the U.S., except that it is printed on oversized paper so as to make the headlines (and the bare breasts) only bigger and bolder. Its charismatic and controversial editor, Kai Diekmann, is proud to have instituted a program he calls the reader-reporters, setting up a phone number, 14-14, to which the public can send pictures they take, often on mobile phones. The paper pays €500 for a photo it uses in its national edition. Bild receives thousands of pictures every day. Thus Diekmann has turned Germany into a nation of paparazzi: No celebrity or funny cat is safe. A few years ago, I showed Diekmann my Flip video camera, and soon he was selling thousands of similar, Bild-branded devices that were set up to send footage straight to the paper. He equipped Germans to make anything they see public.

Now, talking with his editors, I warned of the unintended consequences of the media and government campaign against Street View. If Google can be pressured not to take pictures of public views from public streets, can't Bild's journalists be told not to? What of Bild's reader-reporters? A hush came over the conference room.

The night before, in Munich, I had joined in a public debate over publicness with Wolfgang Blau, editor-in-chief of Zeit Online. In the discussion, a man in the crowd said he didn't like seeing his picture included in crowd shots that other people in the room were posting to the net. That man said he hadn't given permission. If he got his way, I said, everyone else in the room would be prevented from taking and sharing pictures of the event. Would his prohibition next extend to what people said and heard and wanted to share? That impinges on the free-speech rights of everyone else. The ability of people outside the room to follow what was happening there—and comment on it, challenging me, adding ideas and information—would also be restricted. The public record of the event would be limited. The publicness of this event was an asset, and if that man succeeded in preventing others from sharing what happened there he would have robbed us all. What's public is owned by us, the public. It does not belong to one member of that public over another. It does not belong to government. It is society's asset—it's ours. "The streets belong to everyone, and that means Google, too," Financial Times Deutschland said in an editorial. If that man's definition of privacy were to be applied broadly, the paper said, streets "would have to be declared private property. The public sphere would disappear."[11]

In every crowd in Germany, I find people who simply distrust Google because it is too big or makes too much money—or is American. "What makes Google, with its funny name and friendly logo, a monstrous player in the online world, is its desire for omnipotence," Berliner Zeitung has editorialized. "The search engine has become a virtual and omnipresent world player."[12] I hear others complain, "I don't want Google making money on me." This apparent hostility is not consistent with online usage statistics. In the U.S., Google's search engine has about 65 percent market share; in Germany, it has an amazing 93 percent.[13] It would seem that in their attacks on Google, government and media and the loudest complainers are out of sync with German users. Nonetheless, I see a steady barrage of attacks against not only Street View but also Google's Gmail and Buzz, not to mention Facebook (which was taken to court over the privacy implications of putting its "Like" button on sites[14]). German officials and German media may be the most relentless in the world on issues of privacy.

The German Paradox

On a trip to Munich in 2009, I visited the sauna in my hotel. Germans love their saunas. I, too, got hooked. To research this book—I swear, it was for the sake of work—I later visited neighborhood spas in Berlin and Therme Erding, a large water park in Munich's suburbs billed as the world's largest sauna wonderland, with more than a dozen choices of saunas and steam rooms and a giant, warm wading pool with water jets here and there and a bar attached. Germans by the hundreds sweat and shower and lounge around. It's not in the least bit sexual. Spas are a matter of robust health, in the Germans' view. They enshrine this belief under the initials FKK—*Freikörperkultur,* or free body culture. That means nude. And coed. At all these facilities, the schwitzing, swimming, soaking Germans, male and female, hang together in their altogether. Which made me realize that the Germans care deeply about the privacy of everything . . . except their private parts.

That's more than a punch line. Behind that observation lies a lesson and a question for us all: Why is the private private? And why is the public public? As we debate privacy and publicness, we would do well to re-examine our cultural conventions to see what they say about us and our assumptions regarding privacy. We should examine these norms as a society and as individuals, as I will do for myself. In the U.S., nothing is more private than our private parts. You'll get arrested for exposing them in public. In Germany, they say, "What's the big deal? We all have them." I say that's a far more mature attitude than Americans' puritanical prudery. In the U.S., personal finances are probably the second most guarded secret people have after their health information. In Norway and Finland, citizens' taxes and income are published openly. In secretive Switzerland, two politicians enraged their opponents when they dared reveal their own income and taxes.[15] In the U.S., we reveal the identities of people arrested for crimes, putting their photos online and subjecting them to "perp walks" in front of press cameras, but in Germany, published pictures of accused criminals have the eyes covered with those pixel boards Google Man used. In parts of the Middle East and Germany, block walls surround homes to ensure security and privacy. In the Netherlands, convention has it that one should leave one's curtains open, no matter what happens behind them. But a Norwegian told me that in nearby Belgium, a neighbor called the police on a foreigner walking around her own home in underwear with the drapes parted. (I should note that when I used Google Street View to look for an illustration of the Dutch open-curtains policy, I could not find a single window that wasn't draped. Perhaps the neighbors had been warned that Google Man was coming that day.)

When I gave a talk on privacy and publicness at the re:publica 2010 blogging conference in Berlin, the response amazed me.[16] Coverage landed on the front pages of three major newspapers, in the country's two newsmagazines, and on TV. Other newspapers felt compelled to rebut me and start debates about my views. I'd hit a hot button, for sure. I believe the nerve I touched is a nagging fear Germans harbor that their heritage is coming into fundamental conflict with internet culture—with the future. When I spoke about this idea with a group of editors from Die Zeit, a leading journal of reporting and opinion, one of them conceded that I was describing the German culture of privacy correctly. Then he said that his own children did not operate under his rules—Germans' rules—but instead under the internet's. His children were more public. We need to ask, then, whether internet culture will come to supersede local culture. Perhaps what we think of as youth culture—sharing so much on Facebook—is a preview of the society developing around all of us. Will Facebook's norms and Google's mores start to take on the force of a global culture?

The movement toward publicness has a long way to go in Germany. Blogging—and the open sharing of lives and opinions it facilitates—has not taken off there the way it has in many other countries. I asked the two thousand bloggers at re:publica whether I was standing before all the nation's bloggers. "Half!" one of them shouted to knowing laughter. In America, there are millions. Some of my friends say that Germans don't like to share their lives and opinions; they don't even tell one another for whom they cast their votes. Then again, on endless prime-time TV shows, I see Germans sitting on camera sharing opinions aplenty.

When I wrote about this German paradox on my blog, a commenter, Tilmann Hanitzsch, offered an intriguing explanation for why his countrymen are less likely to open up: "We lack a culture of sharing our knowledge," he writes. "We have an antisocial attitude to consider each and every bit of our knowledge as a competitive advantage best kept to ourselves. And we mistrust the fools giving it away for free. . . . The push-button conditioning I grew up with: Have a problem? Don't expose it—somebody will use it against you! Had a success? Keep quiet—it will cause envy! . . . Made a mistake? How embarrassing. Talk about it? Good lord, no! Consequently, we're not only entitled to our own mistakes, we're conditioned to make the same mistakes" others have made.[17]

I played that notion back in Berlin, and many Germans I spoke with supported Hanitzsch's thesis. They told me that Germans have a problem with making mistakes in the open. The Germans I spoke with—internet people—envy American entrepreneurs, who often brag about their failures, viewing them as a public badge of lessons learned. The closed, industrial economy that made Germany such a modern success is being supplanted by the open, digital economy. In that change, my German friends worry, their fear of failure may leave them at a strategic disadvantage.

The notion of the beta is so antithetical to the soul of government regulators in Germany and Europe that privacy czars of seven European nations plus Canada, Israel, and New Zealand sent a letter to Google in April 2010 that not only complained about Street View and privacy missteps with Google's Gmail and its Twitter-equivalent, Buzz. The officials also advised Google against putting out products as betas before they could be perfected.[18] These bureaucrats, conditioned to avoiding public failure, apparently could not imagine Google's motives in willingly making mistakes in public. There the culture clash over publicness comes to life. Publicness is about more than sharing your breakfast on Twitter, your opinions in a blog, or your private parts in the sauna. It also is a window on a society's attitudes toward change and risk, progress and innovation, success and failure.

My Public Parts

I took a liking to the sauna. After my talk in Berlin, I invited the audience to come to the spa next door so we could continue the discussion in the heat, in public, and naked. Four guys took me up on the offer. We sat, sweating, overlooking the Spree River as we talked about our cultural contrasts. One of them blogged about it and reported our discussion.[19]

A few months earlier, in the small Swiss town of Davos, where the World Economic Forum meets each year, I managed to find the one open sauna, set in a log blockhouse outside a hotel. I was getting ready to join a bunch of sweaty Russians, having just taken a shower, when the outside door opened. A woman shrieked, and the door slammed. I heard the man with her—her husband, I assumed—reassure her that this is the way it's supposed to be: men and women, naked and mixed, quite normal. I went into the sauna. The couple soon joined us. She sat, stiff as a church lady, staring straight ahead, looking no one in the eye, sealed tight in her bath towel. After fifteen minutes, I left, refreshed. Later, I blogged about the moment, smugly amused that this woman—whom I assumed was an uptight American—had to learn about European saunas the way I did. Then, on Facebook, she found me. Jasmine Boussem introduced herself as the woman in the sauna. Turns out she isn't American, she's French. The man wasn't her husband but a sauna-loving colleague. The next year at Davos—at a dinner, not in the sauna—Boussem explained why she'd yelped at the door. She knew who I was, having read my blog for some time before she saw me, all of me. She just didn't know what to say. As another German friend explained to me later, saunas are oddly anonymous. Yes, you're naked. But chances are no one knows who you are. You're naked, but you're not.

I learned something about myself in those German spas. I found it was, indeed, no big deal to be naked in front of men and women, even people I knew. As an American, I'd grown up thinking—or just assuming—that nudity in mixed company would be embarrassing, tacky, wrong. But in German saunas, I took on the cultural coloration of my surroundings. I was neither humiliated nor terribly self-conscious. It forced me to examine my own shifting boundary between private and public.

Perhaps that is what has happened to me living online. I have found it easier and easier to live publicly. Granted, it is more comfortable for me to be public than for others. I am a privileged, white, American male. As a writer, I have experience with the consequences of speaking to the crowd. I am not, say, a gay man living under a sexually and religiously repressive African regime who could be imprisoned or killed for being public. My decisions about how public I am would not work for others. I am not suggesting that I am setting an example of publicness. I'm using myself just as one example. These are the decisions I have made. Even in intimate details of my life, I now default to public. So far, I have no regrets.

In September 2009, I had surgery to remove my prostate and its cancer. I emerged, as all such patients do—at least temporarily—incontinent and impotent. I shared most every detail of the experience on my blog and, as as result, in print and on TV and radio.[20]

Any privacy advocate will tell you that there's nothing more private than one's health information. And I'll confess that it's not easy, as a man, to talk about my penis—especially when it doesn't work. But I chose to be public about my condition for good reasons. I had blogged about another of my ailments, atrial fibrillation—an occasionally irregular heartbeat—which I acquired indirectly after 9/11 (inhaling the dust of destruction led to pneumonia, which led to a lung-function test, which used a medication that set my heart a-fluttering the first of too many times).[21] When I blogged about my heart, readers came forward with valuable information and recommendations as well as support.

When I received my cancer diagnosis, my reflex was to go to the blog and talk about it. I had to wait. Our son was away that summer, and I certainly didn't want him to learn about my cancer in a tweet. Once Jake returned, I told him and our daughter, Julia, and the rest of our family. And then I blogged, "I have cancer, prostate cancer." [22] I talked about my reaction—calmer than I'd have expected, given that I'd long been phobic about the disease—and the choices of treatment (surgery, radiation of various sorts, or "watchful waiting"), and my decision, with my wife, Tammy, to get robotic surgery. I also wanted to acknowledge how lucky I was. "If you're going to get it, this is the one to get," the doctor told me, and he was right. I've suffered none of the chemical, radioactive, surgical torture so many brave cancer patients with worse forms must bear. My tumor was small. If "benign" is the most beautiful word in the English language, "contained" is the second best. I am fortunate.

Instantly, comments poured in—there've been 345 on that post to date—with more on Twitter and Facebook, offering support, information, advice, and readers' own stories. After surgery, I continued to blog the experience, culminating under the headline "The penis post," in which I warned readers that I was about to reveal TMI (too much information) should they wish to look away. I wrote about the giant, surgeon-operated robot that had taken out my prostate (my wife learned that patients are put under early because they tend to freak watching the big machine and its arms set up over them); about the catheter I had endured for ten days and its removal (my hosectomy, I called it); about the Baby Huey diaper I'd had to wear (no need, really) and then the pads that followed (I was lucky to give them up after three months); about my lack of an erection, despite Viagra and Cialis (oh, how I resent their warnings of four-hour erections); about doctors prescribing masturbation and penis pumps and the strange experience of having faint, internal orgasms with no external evidence. "We men have complicated relationships with our penises," I blogged. "We follow them (that's why they're in front). They tell us what we like. They have minds of their own. We anthropomorphize them; some give them names (I don't; it's just 'it'). So when I see mine looking like an emaciated, depressed, shrunken old man in a hospital bed, well, it's hard not to empathize." [23] I warned you: TMI.

The experience of surgery and recovery held few surprises for me thanks to the patients who had gone before. They left comments on my blog with frank and open advice about what to expect—exposing intimate details and feelings that no doctor's pamphlet would have given me, advice that carried more credibility because it came from friends and fellow patients. Only because I had announced my condition did I find out that a friend had had the operation ten years before. Andrew Tyndall, who analyzes TV news in the Tyndall Report,[24] sent me email sharing blunt and personal details about what I should expect, from operation to recovery to the remainders of one's sex life. Because I said I was sharing my saga on my blog for the sake of those who'd follow, Tyndall also came online and left a comment that revealed everything he'd told me in the email—and more—in public, "as a contribution to the ongoing Google value of this discussion." He wrote that though some would see my talk about men following our penises as glib, he thought it was profound. He waxed at once poetic and practical about what life becomes when your body—your instincts, emotions, and hormones—no longer lead but follow. "After surgery," he advised, sharing the advice of another patient who had preceded us both, "you will make love like a woman."[25] It's hard to explain what this experience is like, but my friend had the courage to try.

The discussion that ensued was as remarkable as Tyndall's advice. Another patient shared his intimate details but did so, understandably, using a pseudonym. He returned the next day, saying that if we were speaking publicly, under our own names, so would he. Then came Francine Hardaway, also writing under her name, who said: "I am the widow of a doctor who died of prostate cancer twelve years ago. He was so scared about becoming incontinent and impotent that he waited too long to have the surgery; the tumor had already escaped the prostate. He was living the high life of a divorcee and didn't want it compromised." She begged him to get treatment, telling him that it was their love that mattered most. He made her get married before he had surgery. But it was too late. The disease too soon killed Dr. Gerry Kaplan.[26] What makes me gratified—relieved, in fact—that I'd shared my story is that it inspired more discussion and spurred more readers to get screened for the disease. My publicness was worthwhile.

Six months after surgery, I appeared on Howard Stern's radio show. Long a fan of the most frank and open entertainer alive—the king of all media and the prince of publicness—I had called in occasionally to talk about gadgets or technology or Stern's First Amendment fight with the Federal Communications Commission—nothing sexy. I was in the studio that day because Stern's computer adviser from IBM, Jeff Schick, wanted to show me the show's impressive technology setup. Schick and I had sparred in phone calls about using Google services and Apple products vs. IBM's. As I said, nothing sexy.

After a few minutes of nerd talk, as much as he could bear, Stern asked what I was up to next. I told him about this book and said I should ask him a few questions, given his boldly public life. "Yeah, sure," he said, "what do you want to know?" We discussed his fame. But Stern's publicness is about more than fame. His professional epiphany, which he portrayed in his book and movie *Private Parts,* was to reveal his personal life on the air, to be open with his fans as a way to connect with them. "I suppose if it affected some of the people I loved or love, yeah, I have regrets," Stern said. "But all in all I'm glad I did it the way I did it. I always wanted to be open and honest on the air about my life and bring people into it."

I wanted to know whether Stern practiced publicness as a career decision or as an ethic, something he believed in. "I think we're a little too uptight, to tell you the truth," he said. "I think that people should talk about their sexuality and stuff like that. If you're uncomfortable, then they shouldn't do it." But he believes that his frankness set an example for others. "I think it opened up a lot of people to talking. I even think that talking about my masturbation and about lesbianism and all this stuff made it more acceptable to guys who were uptight. . . . It kind of loosened them up and said, 'What's the big deal?'" Talking about his reputedly small penis and his masturbation and sex life, not to mention his humiliations of childhood, Stern tore down just about the last taboos one could imagine—too many taboos, according to his critics and the

FCC. But after revealing his intimate secrets, Stern did not run and hide in shame. His world did not implode. No matter what the FCC did and no matter how high its fines against him, his audience only grew.

I told Stern that he had made me feel freer—even inspired me—to talk about my prostate cancer and to have no fear of saying the word "penis" out loud. Stern being Stern, he jumped right to the key question no one else would ask: "Are you getting it up now?" No, I said. "Shit," he replied. Halfway through a long, on-air discussion, I lamented, "Here's my moment to get on the *Howard Stern Show*, and I talk about my small dick. But what better place to do it?" I managed to make even the Stern crew groan as I answered their questions on details of the surgery and recovery, leaving Stern speechless. "You fucking shut me up," he said after some dead air. The next day, Stern told his audience he had obsessed on the subject overnight, and he decreed, "Every male on the planet should be donating money to prostate cancer research." That's the obvious advantage of going public: attention. We use media—now that each of us wields media—to find allies, inspire action, organize movements, change priorities, influence policy, and raise money. We use our publicness to rally others around our agenda. We also use it to find others in our boat. I've used it for all those reasons, and I am glad I did. I have no regrets.

My Private Parts

After talking about my penis on the internet, on radio, in speeches, and even on TV, you might wonder whether I have anything left to myself. I do. I have a private life. I'm not saying that everything must be public, only those things that have a reason to be. I said above that we—all of us and each of us—should re-examine our assumptions and norms about privacy and publicness. I've just told you about my publicness. Now I'll draw the lines around my privacy.

My starting point: I want to be as careful as possible not to bring others into my glass house: my family, especially, and also my colleagues and friends. In the early days of the web, a popular T-shirt warned, "I'm blogging this." I ask before tweeting what I hear. But I'm aware that no matter how hard I work to separate my publicness from others' privacy, there is publicness by association. When I talk about my prostate surgery, one inevitably gets a glimpse into my bedroom, not to mention my children's DNA. My children have to live with a public dad who not only blogs but makes YouTube videos. I made one video about theatrically reboxing and returning my iPad because I didn't see much use for it and my son thought it over the top. (All parents embarrass their children. I'm undoubtedly worse.)

As to my own privacy: Of course, I don't want my credit-card numbers and passwords to be public for fear of the theft of my identity and assets—but that's a matter of crime more than social convention. I don't want my emails to be public, though don't we all write them cognizant of the possibility—the back-of-mind fear—that they could be shared? A friend of mine recently had reason to fear that all his emails would be unveiled (it didn't happen). After some justifiable panic, he pondered life completely in the open and no longer dreaded it. I don't want my calendar to be public, as there are already too many demands on my time— too many requests for "just five minutes" that add up to more than a day.

I do still get hinky talking about my income and assets. But because I work as a professor in a public university with a collective-bargaining contract, it would take little effort to learn my salary ($90,000+). I have revealed how much I make from blog advertising ($5,000 to $13,000 a year and heading south). The earnings from my first book are a badly kept secret (about $400,000 over three years). You can try to hire me to give a speech and get my price (from nothing for talks that serve my journalistic and academic missions up to $45,000 for a faraway corporate event). You can try to add it up. I just won't add it up for you. Why? Even I'm not sure. So what if you saw my tax return? It's not as if I'm a lottery winner and long-lost cousins will start lining up for new cars. I make a good living, but I work in a city filled with investment bankers who make obscenely good livings. It's not as if I'd be bragging. No, my reluctance is cultural. We Americans may show off what we buy with our earnings, but we don't like to talk about the earnings, do we? If I did, I'd be setting myself apart and people would wonder why. If I lived in Finland, would I be freaked seeing my income published? Probably not, as I'd be looking up others' finances the way we and our neighbors like to find out what one another's homes sold for. When it comes to money, I live by cultural conventions. That's my choice, and I have the controls to exercise it. I'm not 100 percent public.

What else? I don't particularly want you watching as I browse the web. I won't deny seeing porn; find me the honest man who could prove pure. Now that that's out of the way, why wouldn't I share my browsing history? The problem is context: You may draw unwarranted conclusions about me that I'm not able to see and correct or explain. When I got my cancer diagnosis, I went on a binge of internet research and saved many pages to my bookmarking account on Delicious.com, where normally I store boring links to work-related sites. I forgot that a coworker was watching my bookmarks for a university research project. He saw the sudden trend in pages I'd tagged "prostate," and he guessed the reason, eventually asking with concern. I just as easily could have been research-ing the topic for a family member, so his deduction might have been wrong. My colleague could have blurted out something about the news in front of others before I was ready to tell them. In the end, it wouldn't have mattered. I shared the news anyway. But I might have been wiser to have used the controls Delicious affords to mark a bookmark as private, at least until I'd told our children and my employer about my diagnosis.

Though I may not want you staring over my shoulder as I browse, I have no objection to web sites tracking my behavior with cookies, the infobits that let servers know what I've seen. Privacy advocates and some in media—most notably and oddly, the usually probusiness Wall Street Journal[27]—paint cookies as trackers performing secret surveillance, but I see little harm and some benefit in getting more relevant content and advertising. More on cookies later.

What about tracking my movements in the real world? The German Green politician Malte Spitz sued Deutsche Telekom to find all the data it had gathered about his location from his mobile phone. Over five months, it collected 35,000 data points. He made all this public and the news site Zeit Online turned it into an interactive map: Where's

Herr Spitz been?[28] Apple, it turns out, was saving our locations on our iPhones. After a fuss, Apple said it would restrict the practice but it should have told users about the data and let us control it. The E-ZPass automated highway toll system knows much about where I am. Do I have anything to be ashamed about in my comings and goings? Not much. A few too many visits to the local Chipotle, perhaps. My problem isn't so much that the information is collected by technology but that it could be subpoenaed for use against me by government or other foes. Later, I'll examine whether it's better to protect one's privacy by controlling the gathering or the use of such data.

What's left of my privacy? I would find it slightly embarrassing for you to see my iTunes playlist, filled as it is with blathering podcasts, pretentious public-radio shows, show tunes, and torch singers from Joni Mitchell to Norah Jones. I would find it humiliating if my early lovers revealed my learning curve in bed. I won't tell you what I think—especially if it's unflattering—of some of the people with whom I have done business. It would not be in my self-interest.

Those not-insignificant exceptions aside, I'm fine with making much of the rest of what I do, say, or think public. But I don't make it all public—every rumination or speculation, wish or wonder—because, frankly, who could give a damn? I have no desire to be seen as the internet exhibitionist. Some will say I already am one—blogging itself, and more so blogging about one's private parts, can earn one such a reputation. So when I go public, I try to ask why. What's the value? Will I just be creating more chatter and clutter? Will I add to knowledge? What's the benefit?

The Benefits of Publicness

East of Hollywood, most Americans were brought up with a social stigma against being too public: Don't be a show-off, a braggart, a narcissist, an exhibitionist. Don't draw too much attention to yourself. Don't set yourself up for a fall. That varies from culture to culture, of course. As an American, I'm well accustomed to public displays of affluence, affection, accomplishment, opinions, taste, cars, and Christmas lights. But even as an American, I was raised not to poke my head up too quickly.

Then again, fame is our culture's drug of choice. Talk shows and reality shows exploit our lust for popularity like pushers. It used to be that most of us didn't have to worry about fame because we'd never get more than our fifteen minutes of it. But now we all have the potential to live in public. Today we ask ourselves—as individuals, companies, and institutions— how public is too public? How public is public enough? To decide that, as most of us must these days, we need to calculate not just the risks of publicness but also the benefits. Here are some of the benefits that I see.

Publicness Builds Relationships

The market values Facebook not according to how many computers it owns but according to how many members it has, how loyal and active they are, how many connections they make, how much the company knows about each of them, and what it can do with all that knowledge. Compare that with Microsoft, which has never had a relationship with me or connected me with others. It has never gotten to know me. Amazon.com's real value is not in its real estate or the inventory in its warehouses—indeed, it tries to keep as little of that on hand as possible—but instead in how it takes my purchases, builds a profile of my interests and tastes, and tries to sell me what I'll want. Amazon.com has also begun telling me what is popular with my Facebook friends. No department store ever did all that for me. If you're hiring employees, LinkedIn lets you see and sort résumés and make connections through people you know to check out candidates and get introductions. I know start-ups that staff entire teams this way. Old-style headhunters instead keep everything—lists, résumés, meetings, relationships—secret because that's what used to make them valuable. One system is open, the other closed.

Companies will soon measure their worth more by the quality of their relationships than by the cost of the things they own (which, in a digital world, is becoming more of a liability than an asset—witness bookstores and their bricks and shelves or the U.S. Post Office and its offices and trucks). Relationships will come to be worth more than corporate secrets (for what is the value of keeping a mediocre dress design under wraps when by sharing it you can learn what customers really want?). Relationships may be more telling about a company's prospects than quarterly income (for relationships build real long-term value and create a true barrier to entry for others). Brands equal relationships. This is what Mark Zuckerberg is saying when he argues that every product and all business will be social. "Get on the bus," he advises.[1]

I remember sitting at the Quadrangle Foursquare conference in New York watching a private-equity man ask founders of YouTube, LinkedIn, and Twitter which powerful conglomerates they'd want to negotiate exclusive distribution deals with. The entrepreneurs cocked their heads at the notion, looking like confused German shepherds. Their users, they told him, already distributed their services; exclusive deals would only make them smaller. Their companies could start and grow inexpensively and quickly because they relied on open platforms. They didn't see themselves as corporations in industries as much as members of ecosystems—made bigger and more efficient through their relationships.

Rishad Tobaccowala, chief strategist for the digital ad company VivaKi, advances Ronald Coase's theory of the firm[2]—that companies perform tasks inside when it is easier and more efficient to do so than out in the market. Tobaccowala argues that now, in a networked economy, it's getting easier to work with outsiders. That changes relationships with customers, suppliers, and even competitors. J. P. Rangaswami, chief scientist of Salesforce.com and an influential economist and technologist online, recalls a lesson Boston University professor N. Venkatraman taught him: "Businesses used to be hierarchies of business units whose assets were called customers and products." Now "they are changing into networks of business units whose assets are called relationships and capabilities."[3] Turning that perspective into an investment strategy, I'd bet money on start-ups that put relationships at their center so they can disrupt old, closed industries (later we'll look at what social car companies and airlines look like; imagine, too, the social store, restaurant, and school). I'd buy the stocks of companies that know me well and play well with others. I'd short the companies that build walls around themselves. In a linked world and a relationship economy, isolation costs too much.

The internet has changed the infrastructure of relationships. Just as we now take it for granted that any piece of information we want is likely a search away, we are coming to rely on the idea that the people we want to meet are a connection away. That realization is at the heart of LinkedIn, which shows you how you are connected to someone else so you can use those connections to get an introduction and make a relationship. When I have taken to my blog or Twitter to discuss problems with a product or service—as I have with cable TV, phones, airlines, cars, and computers— I can be more and more confident that someone from the company—if it is paying attention—or a fellow customer will come forward to help me fix it. A relationship forms. When I've shared my ideas online, I've found people willing to spread or challenge them (or to do business together). More relationships are born. Since joining Facebook, I have reconnected with old friends and even long-lost family. Relationships are reborn.

To make those connections, we must be public and share. To join up with fellow diabetics or vegetarians or libertarians or *Star Trek* fans, we first have to reveal ourselves as members of those groups. It's the same in the digital world as the real one: If you stay in your room all day, you'll never meet anyone and never know whom you've missed. It's Tinker Bell in reverse: Each time you don't share, a relationship loses its wings. That is a tangible loss.

Publicness Disarms Strangers

Publicness challenges the notion of the stranger. Who's a stranger, when, in any given minute on Facebook or Twitter, someone unknown can become known? A couple of developers in the United Kingdom created an iPhone app called the Situationist[4] to combat what they call "media's demonization of strangers." The Situationist has one user walk up to a stranger who's also using the app and perform a task: hug him, ask her for an autograph, or compliment him on his haircut. It's just a cute and gimicky app but it makes the point: A stranger is not a stranger anymore. People we don't know are just fellow citizens or potential friends. But in media, the app's developers say, strangers are "portrayed as potential stalkers and maniacs. No wonder we're not keen on the public good if we're told the public are nutters."

From the 1970s on, Iran was a strange land of strangers I knew only on TV, mostly in the images of angry young men yelling about us for reasons I'll confess I too shallowly understood. Then another medium came to speak for Iran. Blogging, which took off there very early, became popular as a way for Iranian young people to express themselves about poetry, their lives, and politics, too. One of Iran's leading bloggers, a journalist named Sina Motalebi, was arrested in April 2003 because of what he had blogged. Before reporting to the police, he told his blog readers what was happening. "I publicly announced on my blog that I was summoned to the special police office, something I never did before when they summoned and interrogated me repeatedly," he recalled in an email. He predicted on his blog that he was about to be put in jail. While he waited for the taxi to take him to the police, he watched the reaction to that post begin to erupt online. Soon bloggers around the world—I was one of them—wrote that one of our own was behind bars. That reaction "put my interrogators on the back foot, at least for the first interrogation session," Motalebi says. "Since then, I always encourage anybody who has been summoned to a court, or the family members of friends and colleagues who have been arrested, to be public and loud about the arrest and also find a way to pass the information to the detained person." To this day, Motalebi credits his publicness, his relationship with bloggers, and their spreading word of his situation with getting him out of prison and safely out of the country in December 2004. Today he is an editor for the BBC in London. He says he owes "the life I'm living and the freedom I enjoy" to those who brought attention to him. We are Facebook friends.

I have more friends because I am online. It's easy to mock my 2,000-plus pals on Facebook: How can anyone count friends in the thousands when the Dunbar Number decrees that the most quality relationships humans can manage is 150, give or take?[5] How can I follow 1,000 people on Twitter, and why would more follow me? Of course, the people I know on Facebook and LinkedIn, in conferences and the office, in town and around are not all bosom buddies. But I value them and am grateful for the tools that help us connect, even if it is just to answer an occasional question or follow a recommended link or catch up once in a while. Some people take all comers on Facebook, and others befriend only those they'd invite to a party. For me, these tools and the online places they foster have become the town diner: I never know whom I'll find there, what I'm going to hear, and who might become a true friend. That's why I keep coming back, for the relationships.

Publicness Enables Collaboration

The Wikimedia Foundation, which runs Wikipedia, conducted a study in 2009 adding up the time that users of Wikipedia put into editing it. They couldn't track the effort put into research and writing (it might occur away from the computer, even in a library), but they could track the time spent editing. The study ascribed a conservative $10-per-hour labor cost to that time, according to the foundation's executive director, Sue Gardner. To the foundation's amazement, the total effort contributed by Wikipedia editors added up to $700 million in a year, building an asset that by some estimates is worth more than $4 billion. That is real value created by public collaboration.

Granted, Wikipedia is exceptional. It is so successful at collaboration because its users see it as theirs. Can big, for-profit companies operate so communally? Perhaps not, but they can work together with customers; I'll explore some examples later. They can use open-source platforms—see how IBM has embraced and contributed to the Linux operating system.[6] Don Tapscott and Anthony Williams tell the story in *Wikinomics* of the mining company that opened up its geological data to get more minds working on finding pay dirt. Companies may not come to resemble kibbutzim, but they can still open up and collaborate.

Return again to the notion of the beta. When a technology company releases a product as a beta, unfinished and imperfect, that is a public act, which reveals the process of development for all to see. It is necessarily a call for collaboration: "This thing isn't done," the beta label is saying, "so help us finish it." Opening up in such a way gives customers a measure of respect. "You might have better ideas than we do," it says, "so help us improve what we're doing." The beta resets the relationship and improves it. The beta recognizes that it's better to work together than alone. The beta acknowledges that customers can and often should be cocreators.

Take TCHO, a San Francisco chocolate company founded by rocket scientists who left the space shuttle program. "Since a lot of us have tech backgrounds, we adopted a familiar idea: We encouraged our users to help us make the chocolate they wanted, in much the same way software developers engage beta testers," the company says on its web site.[7] Its first official chocolate went through 1,026 iterations before being released as a 1.0. "When one of our chocolates graduates from beta, it means we've integrated your feedback, finished our tweaking, and believe it's ready for general release—which means much bigger batches."

Not long ago, I stood at a whiteboard before executives at a retail company and suggested they could act as a bridge between customers and manufacturers to foster collaboration. The goal is to move the customer up the chain—the design chain, the marketing chain, the service chain—hearing from customers earlier to act on what they say. Imagine if a retailer got customers to design their ideal product and took that design to manufacturers, saying, "If you make it, they will buy it, so we will sell it." Soon after, I sat on a plane next to a maker of briefcases and bags sold by that retailer. I asked him about the idea of a store getting customers to specify the features and design of, say, the ideal road warrior's bag. Would this bag maker make it? Of course, he said. Collaboration in design would give him a product with proven demand, which could increase sales and lower risk. It would also change his brand, making his the company that empowers customers, that turns them into partners. Collaboration isn't just about making nice. It makes economic sense.

The problem historically has been that customers enter the chain only at the end, after it is too late to act on their ideas. Even retailers—who have direct relationships with end users—come into the process too late, receiving products after they're finished. Sure, manufacturers may pay for surveys or focus groups to get input, but those are little more than random bunches of people who have nothing better to do than give opinions about topics they may not care about for $20 plus cookies. That's not collaboration.

I hear companies fret that revealing and discussing things openly will tip off competitors, allowing them to steal good ideas. That's a problem if you think that your value resides only in your product and that secrecy itself is worth protecting. If, instead, your value lies in the quality of your relationships, openness brings benefits. If you are known as the company that collaborates with customers to give them the products they want, you may end up with more loyal customers. If you are a customer, wouldn't you pick the company that makes better products through collaboration, the company that listens?

Publicness Unleashes the Wisdom (and Generosity) of the Crowd

The more we open, gather, analyze, and share our knowledge, the more we all know. Google's engineers found that by tracking search queries for "flu," they could map the spread of the disease around the world ahead of the U.S. Centers for Disease Control and Prevention, helping health-care officials forecast the need for vaccine and treatment.[8] If each of us went only to our own doctors to seek information, it would be much more difficult to aggregate, track, and analyze that information. That we ask the same third party, Google—and can do so anonymously—adds up to public knowledge. For that reason, Google cofounder Larry Page told European regulators they should not be too quick to erase search data out of privacy concerns. To map trends and anomalies over time might allow Google and health officials to plot and predict the course of the next pandemic. "That could possibly save a third of the world population," Page claimed.[9] At the Personal Democracy Forum in New York in 2010, U.S. Chief Technology Officer Aneesh Chopra told how the government's releasing hospital data in an open standard allowed Microsoft's search engine, Bing, to plot that information on its maps so users could find not only the nearest but also the best hospital to treat the flu.

As I've said, there's nothing more private than our health information. But why? What's the harm of sharing that data? There are many concerns. One fear is that insurance companies will reject us. But they already force us to sign over our medical histories. That is why the so-called Obamacare outlawed rejecting customers due to preexisting conditions. The law deals with the problem by restricting the use, not the flow of information. Another fear is that we won't get hired because of a medical problem. That, too, is society's problem to solve. If employers may not discriminate on the basis of age, gender, race, religion, or disability, should they also be forbidden from discriminating on the basis of health? As many of us get our DNA mapped, will we need to forbid discrimination on the basis of genes? A larger fear of sharing health information is the stigma associated with illness. That stigma is most certainly society's problem. Why should anyone be ashamed of being sick?

Consider a condition that is, by its nature, visible and thus public and carries its own stigma: obesity. There's no hiding fat. Many countries now face crises of obesity and are grappling with its health risks and costs. New York's mayor, Michael Bloomberg, ordered restaurant chains in the city to post the caloric content of every item (it has changed the way I order a nosh at Starbucks, I can tell you). By tracking public data, students at my journalism school working with a colleague dug deeper into factors that contribute to the problem of obesity in poor neighborhoods, where the high cost and lack of availability of fresh, healthful food—tied to the low cost and easy availability of high-calorie fast food—contribute to obesity and diabetes. Open information about the problem will help us address it.

Rather than refusing to talk about weight because we think it is embarrassing for the overweight person, isn't it better—isn't it healthier—to encourage people to discuss their problems openly and to encourage others to offer solutions and support? A young star reporter at The New York Times, Brian Stelter, wanted to lose weight, so he tweeted everything he ate, reporting his diet publicly to pressure himself. That also allowed others to support and pressure him.[10] Stelter confessed in The Times that he had problems at first telling even Twitter the truth and fell off the social wagon, not fessing up to a late-night slice of pizza. Then he found an audience. "We'll be your support group," said one reader. His brother started a Twitter diet alongside him. Friends told Stelter he was changing their habits by example. His disclosure became an act of generosity, helping others. He came to want to share. Exposing fast food's fat and calories became his cause. "Monday, started w/McD's, cinnamon melts and hash browns, 600 cals/44% of day's fat—awful, and made me feel ill," he tweeted. He even summoned the courage to buy a Wi-Fi scale that tweets one's weight automatically, for all to see (no lying possible). Stelter lost ninety pounds on the Twitter diet and tweeted: "I haven't fit into jeans in give or take ten years . . . Jeans shopping for the second weekend in a row. And I must say, it feels great."[11]

There's a twist in Stelter's story: He started his career writing the definitive blog covering the cable news industry, called CableNewser. He wrote it anonymously because he was only nineteen years old. If his industry audience had known he was a mere teen toiling in a dorm room, they likely wouldn't have paid him much attention. The Times outed his age in a page-one feature. He sold his blog to another company, expanded it to cover broadcast news, and when he graduated, he got his job at The Times, where his byline is appearing on page one with regularity. Stelter found shelter in anonymity and then benefit in publicness.

Sir Tim Berners-Lee, the inventor of the web, said at a Google conference in London in 2010 that the data we make public become yet more valuable as they mix with other data. We can find new correlations, trends, and cause and effect in the aggregation.[12] He argued that in government and elsewhere, we should make data public by default, using standards that enable such analysis. At the event, privacy advocate Shami Chakrabarti, director of the U.K.'s National Council for Civil Liberties, attacked Berners-Lee. She bristled at the idea of massive databases, jumping to the conclusion that they would violate privacy. Berners-Lee countered that after eliminating data that hold personal information, there is still an untold wealth of knowledge to be found in what remains, and we should not lose the opportunity it affords us. Mining that data may become the gold rush of our age.

A start-up called Kaggle facilitates contests to analyze open data. In one, government agencies in Australia put up data on traffic patterns and challenged the 364 teams that entered to find better ways to predict delays, enticing them with a $10,000 prize. The winning team's analysis found, counterintuitively, that traffic jams can propagate both ways— that is, a slowdown behind you can end up catching up with you. Also on Kaggle, Ford offered $950 to come up with an algorithm that takes various data points—phone calls, conversations, eating, fatigue—to help determine which drivers are distracted. The Heritage Health Prize, which Kaggle administers, offered $3 million to the team that can best predict who will be hospitalized in the next year. These projects are made possible with open data.[13]

Just look at what we have created with shared data so far: Wikipedia; Google search, which is built on using our links and clicks to learn which sites are most relevant; Wolfram|Alpha, which tries to make sense of more complex data; Google Maps and open-source mapping projects, which collect our photos and annotations; review sites such as Trip-Advisor for travel, Yelp for restaurants, and Rotten Tomatoes for movies; PatientsLikeMe, where patients share details about their medications and treatments; Twitter, Facebook, and Quora, which give us a place to ask questions and get answers; Ushahidi and SeeClickFix, which let people report anything from graffiti to disasters around them . . . the list can and will go on and on.

Publicness Defuses the Myth of Perfection

Thanks to the practical realities of our industrial economy—the efficiencies of mass production, distribution, marketing, and media—we are saddled with a myth of perfection in modern society. A one-size-fits-all "perfect" product that takes a long time to design and produce is sold to a large market. Its manufacturer cannot have it perceived otherwise. There are no second chances on the assembly line. The distribution chain invests in large quantities of the product and cannot afford for it to be flawed. Mass marketing is spent to convince customers that the product is ideal. So perfection becomes our standard, or at least our presumption: our shared myth.

But perfection is a delusion at best, a lie at worst. It is unattainable. The claim of perfection supports priesthoods with closed orthodoxies who define standards for all in fashion, publishing, education, and entertainment. Perfection inflates expectations and inevitably disappoints (every car eventually breaks). Perfection discourages risk and innovation, openness and invention. Perfection is expensive. And the quest for perfection leads only to failure. After all, nothing and no one is perfect.

By operating in public, warts and all, we no longer hold ourselves to the ideal of perfection. By rejecting perfection as a promise, we are free to make what we do ever better. We are never done, never satisfied, always seeking ways to improve by working in public. *"Le mieux est l'ennemi du bien,"* said Voltaire: The best is the enemy of the good. The best is also the enemy of the better. Striving for perfection complicates and delays creation. In technology, we call this insidious process "feature creep"— adding one more gewgaw to get one step closer to the ideal before release. The cure is the public beta: Just put it out there to see what it needs.

The tyranny of perfection permeates the rest of society and our lives.

In our schools, we teach students there is one and only one right answer to every question. Then we add the questions together in tests and teach to those tests, expecting students to spit back what we feed them. We call that achievement. We should instead be encouraging experimentation, rewarding challenges to our accepted wisdom, and designing schools around learning through failure.

The expectation of perfection hampers government. A few years ago, I spoke about Googley government with five hundred federal webmasters in Washington, D.C. Those geeky civil servants are among our best hopes for innovation in government. But they and their bureaucrat bosses live in dread of mistakes. They know that one misstep can bring the disapproval of their bigger bosses—politicians—and of media and constituents. Public servants need a license to fail so they can try things in public, imperfect and incomplete, and collaborate with us all. The webmasters cheered at the suggestion. But I found few who were optimistic enough to believe that day will come. We still inhabit a culture that wants heads on platters when mistakes are made, especially in politics. Search Google for "politician resigns," and you'll find a parade of ignominy.

Now search Google for "CEO apologizes," and you'll find a pile of crow bones on the plate with Toyota, BP, Citigroup, Chrysler, and even NPR at the table. When Michael Dell returned to Dell and Howard Schultz came back to Starbucks to fix their respective companies, each was open about their problems. Dell had quality, customer service, and reputational issues, which I recounted (and, to some extent, caused) in my blog and last book.[14] He instituted the means to listen to customers' complaints and ideas and act on them. Starbucks, Schultz believed, had watered down its experience.[15] He went so far as to close stores while baristas were retaught how to make a cup of coffee. Groupon CEO Andrew Mason at first defended controversial 2011 Super Bowl commercials that seemed to make fun of suffering Tibetans and dying whales until it became clear that the public didn't appreciate the jokes. He apologized: "We've listened to your feedback, and since we don't see the point in continuing to anger people, we're pulling the ads."[16] Those CEOs trusted their customers. They learned that responsiveness beats defensiveness. Confession is as good for the PR strategy as it is for the soul. "We believe that disclosure of oneself to others is a moral good in itself," Richard Sennett says in *The Fall of Public Man*.[17]

Our myth of perfection, I suspect, also affects our personal lives: our romances and marriages and our relationships as parents and children. How many wives are caught trying to fix their husbands' failings—and failing? When should we push our children to succeed, and when are we holding them up to some false and unattainable standard?

Granted, the problem with my attack on perfection is that it could lead to lower standards, to settling too soon, to the scourge of being just good enough. But I believe publicness and pride will save us from that mediocre fate. Even if imperfect, no one wants to seem shoddy in public.

Publicness Neutralizes Stigmas

The flip side of perfection is the taboo. Publicness can disarm that, too. Perhaps the best illustration of the power of publicness is how it has given gays the keys to leave their closets. When gay men and lesbian women for so long were forced to hide their sexuality, it was a commentary not on their mores but on society's. Secrecy did not give gays control over their lives; it granted control to the bigots who forced their norms on others. The solution for gays was to come out, to be public, to show pride, to gather in solidarity and in strength, and to defy society to disapprove. Such publicness says to the world, "Yes, here's who I am. So now what?" Or as we say in my homeland, New Jersey, "Ya gotta problem w'dat?" Publicness is a dare. It takes courage to face one's critics and haters directly, to confront the fear of what people will say about you.

To be clear, I am not suggesting that men and women should be forced from their closets. (I might make an exception for particularly vitriolic antigay public figures who turn out to be gay—though what is being exposed then is not their sexuality but their hypocrisy.) Revealing one's private life should be one's own choice. But when one does make that brave decision, it can change minds. "Since privacy cloaks norm violations from society's view," says Daniel J. Solove, "privacy can serve to retard the process of changing norms."[18] That is, secrecy prevents change; publicness accelerates it.

Society has taboos for a reason, of course. Fearing reprisal from our community is one force that underpins civil society as it stops us from doing wrong. Or does it? If you had no reason to fear what people would think of you if you did something you know is wrong, or if you knew you could do the deed in secret, would you steal your neighbor's newspaper (or spouse)? Or does your moral compass have its own true north? Is the worry about what others will think enough to make you dye your hair, change your accent, not wear the outfit you like even if it is out of style, or say something you don't believe? What's insidious about the fear of what others will say is that you rarely hear them say it. You imagine what they'd say. You imagine they care that much about you. The fragility of our own egos gets the better of us.

Living in public not only shows that we have little to hide; it shows we have little to fear. "What would privacy be like if it weren't connected to shame?" asks *The Cluetrain Manifesto*.[19] Which reminds blogger Dave Pell of sitting in his child psychiatrist's waiting room at age eleven when a kid from his school, Brad, walked out: "My worst nightmare," Pell says. Now he realizes they were hardly alone, Dave and Brad, in seeing a shrink. What if he had known that then? What if Dave and Brad had been open about seeing their doctor? He asks, "Can shame survive in a world without privacy?"[20]

Publicness Grants Immortality . . . or at Least Credit

Most of us do secretly want to be famous, don't we? Yes, there are degrees of fame and there are limits to what a sane person will do to acquire celebrity. Later, I'll examine oversharing. But for now can we stipulate that attention and credit feel good and often just? Can we agree that the desire for these rewards is part of human nature?

Fame is the ultimate extension of human identity. Animals aren't known for anything—or even if they are, they don't know it. (Did Lassie understand what it meant to be Lassie?) But we humans want attributes attached to our identities. We want to be known for something. We want the hallmark of our lives to live past us, with our reputations and creations as our legacies. To that extent, most of us want a public persona.

I will admit to savoring the precious few moments of demicelebrity I've had. When people say they've read my book, I beam. When I was a columnist for the San Francisco Examiner early in my newspaper career, my face was plastered on news boxes for a month. A few times a stranger recognized me on the street—which was especially wonderful if I happened to be on a date. I am a regular panelist on the podcast *This Week in Google*,[21] and from that I've been amazed to be recognized in Munich and Vancouver airports, in a New Jersey Fuddruckers, and on a New York sidewalk. I can guess what you're thinking right now: Bragging about these moments is rather unbecoming: egotistical, show-offy, yes? You're right. Sorry. I'm just being honest. I like the attention. I'm human.

We believe fame should be earned. Fame at its best is credit—credit for accomplishments, contributions, talents. If we do something just for the attention—especially a charitable act—don't we also believe that the grab for glory cheapens the act? But by claiming our deeds and ideas, we also claim responsibility. We create a public record. Fame can be good.

Immortality is even better. Only by being public can we leave our mark on the world. Hannah Arendt argues that if we are not public, we are the trees that fall in forests that no one hears. To be private, she says, is "to be deprived of the possibility of achieving something more permanent than life itself. . . . Private man does not appear, and therefore it is as though he did not exist."[22]

In our quest for modern immortality, we have a problem with the architecture of our new public realm: Data turn out to be surprisingly impermanent. Have an old floppy disk in the house (if you're not too young to wonder what that is)? Any idea how you'd read it? "Floppy disks won't last as long as the Gutenberg Bible, which has lasted 500 years," says Elizabeth Eisenstein, author of the definitive work on the birth of printing. She worries about preservation. So does Dave Winer, a true pioneer of the net who has had key roles in the start of many online technologies, from blogging to RSS to podcasts. He wonders how we can preserve our digital lives—our blogs, photos, Facebook pages, and legacies—after we are gone.[23] To whom can we entrust them: to family or to one of the dozens of companies[24] now making a business of insuring our legacies (see Entrustet) or perhaps to universities—any of which might also die? "There is no mausoleum," says Eisenstein. There is no library of our lives.

Publicness Organizes Us

Protestors in Iran used Twitter as they demonstrated against the dubious results of the nation's 2009 election. All too quickly and perhaps glibly, their movement was dubbed the Twitter Revolution by many in the digital West—notably by political blogger Andrew Sullivan. He tirelessly blogged the news there by combing through Twitter alerts about it. Twitter, he declared, would henceforth be "the critical tool for organizing the resistance in Iran."[25] Sadly, the idea that Twitter fueled that revolution—like the revolution itself—turned out to be in great measure wishful thinking. We do a lot of that, we internet evangelists: believe that one tool or another will change our world overnight. New York University professor Clay Shirky calls that the "'just-add-internet' hypothesis."[26]

Evgeny Morozov, an editor at Foreign Policy and a fellow at Georgetown University, as well as a respected supplier of ballast to me and my fellow internet triumphalists, argues in Dissent and Prospect magazines that in Iran, Twitter was useful mainly to "a tiny and, most important, extremely untypical segment of the Iranian population," as well as to Westerners who couldn't get news from journalists who had been deported or restricted. He says Twitter "only add[ed] to the noise" and, along with Facebook, gave "Iran's secret services superb platforms for gathering open source intelligence about the future revolutionaries. . . . Once, regimes used torture to get this kind of data; now it's freely available on Facebook."[27] Yes, the tools can be used by either side in a dispute.

In Prospect, Shirky counters, "Because civic life is not just created by the actions of individuals, but by the actions of groups, the spread of mobile phones and internet connectivity will reshape that civic life, changing the way members of the public interact with one another." He adds, "The new circumstances of coordinated public action, I believe, marks an essential change in the civilian part of the 'arms race.'"[28]

Morozov and Shirky are both right. The internet is still new. It is a deep mine filled with unseen potential. It won't be used in all the ways we want. It will surprise us for good and bad. But it would be a mistake to declare one use or impact null just because an early encounter did not meet overblown expectations. I, like Shirky in his book *Here Comes Everybody,* take it as a matter of faith that the tools of the internet will facilitate coming together in new ways: to organize clubs, cults, companies, markets, revolutions, even new forms of governance that exist outside of government—as well as criminal syndicates and terrorist organizations.

Looking forward from Iran, we see how revolutionaries in Tunisia, Egypt, and other countries used Twitter, Facebook, YouTube, and other social tools to organize themselves. In Libya, activists hid coded messages in poetry on a dating site to spread their word.[29] Any tool in the storm. Looking back, earlier revolutionaries used the tools they had at hand. I am writing this section of the book while in Berlin. This morning, I ran on an awe-inspiring route—through the Brandenburg Gate and along Unter den Linden to Alexanderplatz, the hub of the former East Berlin. At the end, I came across an outdoor exhibit about the movement that led to the fall of the Wall and Germany's reunification. There, in one display, were some of the crude printing tools used by dissidents to publish their tracts. Farther along in the chronology was a worn Amiga computer and the dot-matrix printer protestors used to print leaflets and spread their message farther, faster. Was theirs the Amiga revolution? Imagine if that computer could have been tied to the internet along with millions more devices like it. Would dissidents have found one another sooner and realized that they had the critical mass—the safety in numbers— they needed to step out in public and bring down the government and its Wall? Or would the Stasi have hunted them down more efficiently? Would we have had a new Europe sooner or not at all?

Publicness Protects Us

The more public society is, the safer it is. That assertion will be unsettling to many because it summons fears that technology can help any government—including repressive regimes—watch us, use our actions against us, and make us public against our wills. But can we afford to return to the days when government didn't know who flew on our planes and what they carried? Where's the balance?

Search Google News for "Orwell," and you will find a ceaseless flow of fears about government's eyes watching us everywhere. Examples abound: As a legacy of the sectarian terrorism it suffered, London still has CCTV cameras recording, it seems, every step in the city: Big Brother's gaze, it's alleged. There is no authoritative count, but estimates in 2008 ranged from 1.2 to 4.2 million cameras.[30] A single London bus can have as many as sixteen cameras.[31] In its hunt for terrorists after September 11, 2001, The New York Times reported, the U.S. government listened to citizens' otherwise private phone conversations without warrants: Big Brother's ears.[32] Local governments use cameras to catch and ticket drivers who violate traffic laws. Towns in the U.K. use aerial infrared photography to spot poorly insulated homes,[33] posting maps so that one can check on the energy-leakiness of one's neighbors.[34] Media companies track, subpoena, and sue kids and moms for violating copyright law.[35] In 2011, the Obama administration proposed criminalizing unauthorized media streaming as a felony and empowering the FBI to fight it using wiretaps.[36] Following various financial scandals, governments demanded that companies be more public about their businesses in ways that their managers say makes doing their jobs too difficult. Every time a government laptop is reported lost or stolen, we are reminded of how much information government holds on us and how vulnerable it is to exposure and abuse.

"Trade-off" is the word that often punctuates this discussion. We have a choice: We can condone government use of technologies to track bad actors and keep us safe. Or we can forbid governments to do so because we fear they will use it to track us and threaten our privacy and liberty.

Germany's minister of food agriculture, and consumer protection, Ilse Aigner, decreed in 2010 that tying facial recognition to geolocation technology would henceforth be "taboo." [37] On the face of it, that sounds sensible, for the idea does sound sinister. But is it wise to ban a technology before it is even used and understood? Imagine how else such a combination could be beneficial: finding missing children or learning the fate of victims in a disaster such as Hurricane Katrina or the 2011 Japanese earthquake and tsunami. Trade-off. It would be "hilarious to see such a law passed in the U.K., where facial recognition is being used to scan airport crowds for known terrorists," says Eric Schmidt, executive chairman of Google. "It seems to me that it's probably a lot cheaper to scan the crowds for known terrorists than to do this invasive security searching that they have to do for everybody." In 2010, a media hubbub burst out when some travelers objected to full-body scanners that show fuzzy, quasi-naked images of travelers to U.S. Transportation Security Administration officers somewhere else in an airport. If they opted out, passengers were subjected to pat-downs, deemed by some to be too invasive, as they encroached on the groin—because that's where the "underwear bomber," Umar Farouk Abdulmutallab, hid his plastic explosives on Christmas 2009. [38] Knowing that no security at all is not an option, what's your choice: body scans, physical searches, facial recognition via surveillance cameras, more personal data attached to travel records? Trade-offs.

I object to none of the above. But I should reveal this context: I was at the World Trade Center on 9/11, on the last PATH train to arrive under the North Tower just as the first jet hit. I was standing across the street as the second jet exploded and ran away when the South Tower fell. So I may be different from most people. I've set my tolerance for surveillance high. As a society, we are negotiating our mutual limits. We do need limits and controls. Terrorism is an edge case, an extreme that can be used to justify and hide other possibly invasive government actions. But terrorism is also a reality.

In another use of surveillance technology, the town of Riverhead, New York, used Google's satellite images to spot 250 illegal swimming pools in backyards that had not received safety certification. [39] Is it important for us to ensure that our neighbors' pools are safe for the children around them? If pool owners don't come forward, do we want government to find them out? What methods are OK if not Google Earth? Knocking on doors? Getting a plane and flying overhead? Relying on snitching neighbors? Monitoring water usage? Where do safety and privacy cross and clash? The town set, then reset its priorities: A month after the news spread that Riverhead was using Google Earth, the town council voted to stop. That was a political and emotional response. The images are still online for anyone to see. The need to protect children from dangerous pools is still there. The U.S. Centers for Disease Control and Prevention report that in 2007, almost one-third of the deaths of children aged one to four occurred because of drowning.[40] The hubbub over privacy won this round. Should it have? Wouldn't publicness about pools be a wiser course? Did we have our priorities between the risks to privacy and the benefits of publicness set properly?

A History of
the Private and the Public

Fiendish kodakers

The first serious discussion of privacy as a legal right in the United States did not begin until 1890. Advances in technology were the catalysts: In 1888, Kodak introduced its first boxy, portable "snapshot" camera.[1] Freed from the studio, where it had been used to shoot portraits, the camera could now be carried anywhere, taking pictures of anyone. Those images could be seen by everyone thanks to the growth of mass-circulation, illustrated newspapers: the penny press. In the half century leading up to 1890, in the United States alone, their number multiplied from one hundred to nine hundred publications and their total circulation exploded from 800,000 to 8 million.[2]

All photographic hell apparently broke loose. The New York Times reported in 1899 that "kodak fiends" were harassing the good ladies of Newport: "All over the avenue the women are constantly brought face to face with a kodak and snapped," the paper huffed.[3] In 1903, Reggie Vanderbilt, heir to the Vanderbilt fortune, horsewhipped a "yellow kodaker"—that is, a photographer for the sensationalistic popular press—and The Times cheered, arguing that though he was a member of a rich family, this Vanderbilt was not a public figure and should be left alone. The paper noted that President Theodore Roosevelt had been "known to exhibit impatience in discovering designs to kodak him," which was why he briefly outlawed cameras in Washington parks. But The Times argued that, unlike young Reggie Vanderbilt, Roosevelt was "a public and kodakable character."[4]

Even while cameras were still too big to lug out of studios, the possibility of seeing one's portrait published raised questions of privacy the law could not yet settle and would not resolve for many years to come. "The right of a man to control the publication of his own features is a rather delicate point of personal law which has never been sufficiently elucidated," The Times wrote in 1874.[5] In 1897, a bill was introduced in New York to levy a $1,000 fine and a one-year prison term for publishing a portrait without written consent.[6] The Times spoke favorably of the legislation at first,[7] until its editors apparently realized that they could be liable themselves; they eventually labeled it "bad law."[8]

In a 1902 case involving the use of young Abigail Roberson's likeness pasted on flour barrels without her consent,[9] the New York Court of Appeals declared "that there is at present no right of privacy known to the law."[10] Or, as The Times put it, the court "has decided that privacy is one of the trivialities contemplated in the legal maxim, '*De minimis non curat lex.*'"[11] That is, the law does not concern itself with trifles.[12] The court's chief justice, Alton B. Parker, wrote, "The so-called right of privacy is, as the phrase suggests, founded upon the claim that a man has the right to pass through this world, if he wills, without having his picture published, his business enterprises discussed, his successful experiments written up for the benefit of others, or his eccentricities commented upon either in hand-bills, circulars, catalogues, periodicals, or newspapers." Judge Parker reasoned that if this use of an image were deemed an invasion of privacy, the rule could extend to all such uses, positive or negative, in all forms, word or image, in any medium, including newspapers, leading to a likely torrent of litigation.[13] The debate continued a month after Parker's decision, when The Times editorialized that J. Pierpont Morgan had been "beset by 'kodakers' lying in wait" and urged legislators to pass laws curbing "these savage and horrible practices."[14]

Judge Parker later ran for president and came to complain about photographers himself, saying, "I reserve the right to put my hands in my pockets and assume comfortable attitudes without being everlastingly afraid that I shall be snapped by some fellow with a camera." To which Roberson—the woman on the flour barrels whose privacy the judge had dismissed—responded in The Times, "You have no such right. . . . It is not apparent how your likeness in the attitude suggested could be libelous, at least not as long as you kept your hands in your own pockets." [15]

The key moment in the birth of American privacy law came in 1890, when Louis Brandeis, later a Supreme Court justice, and Samuel Warren wrote a Harvard Law Review essay, "The Right to Privacy," in which they proposed a new principle protecting "the right to be let alone." Mind you, publicness is protected in the Bill of Rights—that is the essence of the First Amendment—but there is no article assuring a right to privacy. As the cases above demonstrate, there had been no established legal right to privacy. So Warren and Brandeis had to look for it elsewhere in the law. They came to the question in high dudgeon about the need to protect against gossip in the press. "The press is overstepping in every direction the obvious bounds of propriety and decency," they sniffed. "Gossip is no longer the resource of the idle and of the vicious, but has become a trade, which is pursued with industry as well as effrontery. . . . Each crop of unseemly gossip, thus harvested, becomes the seed of more, and, in direct proportion to its circulation, results in the lowering of social standards and of morality." [16] They went so far as to call unauthorized circulation of people's portraits "the evil invasion of privacy by the newspapers."

It's not known precisely what raised Warren's and Brandeis' hackles. The long-held assumption had been that it was press coverage of Warren's wedding to Mabel Bayard, daughter of Thomas Bayard, a senator from Delaware, presidential candidate, and secretary of state under Grover Cleveland. But that wedding occurred seven years before the Harvard essay and likely wasn't the spark. Other reports blame coverage of the wedding of the Warrens' daughter, but that came later. In a 2008 Michigan State Law Review piece, Amy Gajda analyzes press coverage of the family and finds little interest in Warren; it was his in-laws who fascinated reporters. [17] Gajda proposes that without his marriage, Warren likely would not have been in the papers much and wouldn't have come to care so fervently about corralling the popular press and their kodakers.

Technology Fears

Technology is one thread that ties together outbreaks of fear over privacy through history. In 1890, the cause of trepidation was the camera. Centuries earlier, it was the printing press. Gutenberg's machine, invented between 1440 and 1450, made authors nervous about their own publicness. Being public—having one's thoughts set down permanently and distributed widely under one's name—was new, strange, and frightening. "The paradoxical implications of making private thoughts public were not fully realized until authors began to address an audience composed of silent and solitary readers," says Elizabeth Eisenstein.[18] In 1628, Pilgrim pastor John Robinson worried that by writing books "the Author therin exposeth himself to the censure of all men."[19] Daniel Defoe, author of *Robinson Crusoe,* wrote in 1704—the year after being pilloried for publishing a satirical pamphlet—that "Preaching of Sermons is Speaking to a few of Mankind: Printing of Books is Talking to the whole World." *Gulliver's Travels* author Jonathan Swift, also trying to grapple with the impact of printing, said in 1711, "A Copy of Verses kept in a Cabinet, and only shewn to a few Friends, is like a Virgin much sought after and admired; but when printed and published, is like a common Whore, whom any body may purchase for half a Crown."[20]

Now leap forward to recent history. Alan F. Westin, in his influential 1967 book *Privacy and Freedom,* lists technologies of the 1880s onward that, in his view, threatened privacy. There was the invention of the microphone in the 1870s, the telephone in the 1880s, and the recorder and the camera in the 1890s, each of which could be used by government or the press to spy on citizens. In his own time, Westin found many more technologies to fear: LSD "may greatly affect the individual's daily personal balance between what he keeps private about himself and what he discloses to those around him" and could be used for government surveillance. Westin worried about radio pills, miniature transmitters, and even about fluorescent powders and dyes—not to mention radioactive substances—that could be applied to "hands, shoes, clothing, hair, umbrella, and the like, or can be added to such items as soap, after-shave lotion, and hair tonic" to track the unsuspecting person. Secret miniature cameras, transmitters embedded in hearing aids or shoelaces, infrared film, microminiature microphones the size of match heads, battery-operated tape recorders, hidden "television-eye" monitoring, telephone tapping, "truth measurement" by polygraph tests, personality testing, brain-wave analysis, dossiers of personal data, TV ratings, and the means to steam open envelopes—all those concerned him. He speculated that "invisible magnetic-ink tattoos might be applied (for example, to babies at birth)" and transmitters could be implanted in their bodies. He fretted over "wireless, battery-operated television 'eyes' the size of buttons," not to mention U-2 spy cameras from above as well as scientists' ability to read brain signals. He also warned of the dangers of computers. In 1966, he wrote, there were 30,000 computers in the U.S., 2,600 of them in the federal government. What happens, he asked, if we come to the day when "computers in the field of health will eventually establish total medical profiles on everyone in the country 'from the hour of birth' and updated through life. Each record will be almost instantly accessible to medical personnel." [21] Oh, if only.

Westin listed his worries about technology's impact on privacy almost five decades ago. How many of his fears have come to life? Few if any. That is neither to mock him nor to diminish his warnings, only to put the anxieties technology fosters into context as we grapple with the concerns attached to our contemporary science. The fear of devices continues today: "Information technology is considered a major threat to privacy," Helen Nissenbaum writes in *Privacy in Context*. "It enables pervasive surveillance, massive databases, and lightning-speed distribution of information across the globe." [22] A 2007 book by German Commissioner for Data Protection Peter Scharr, *Das Ende der Privatsphäre* (The End of the Private Sphere), lists more causes for worry: internet attacks, a net that never forgets, radio chips, locator beacons in pockets, biometric identification, DNA identification, and collateral damage to privacy in the war on terror. [23]

Yes, the internet, exponentially faster computers, ever less expensive data storage, ever bigger and more efficient databases, mobile technology, cameras on every corner and in every hand and in the sky above, geographical annotation of information, social networks, and ubiquitous publishing platforms make possible entirely new ways to gather and share information. Bad things could happen. It is prudent and wise to consider those possibilities and guard against the dangers, as our army of privacy advocates does. But those new technologies also present new opportunities, which we could miss if we are too busy building our bunkers. Presses print gossip but also art; Kodak cameras can embarrass yet enlighten; digital cameras power spying as well as grandparents' Skype video calls; orbiting cameras equip spy satellites and Google Earth. What makes technology as frightening as it is exciting is that it is so unknown. In our messy tangle of wires and the frightening sparks shooting among them lays progress. Author Douglas Adams wrote in a 1999 newspaper essay:

> I suppose earlier generations had to sit through all this huffing and puffing with the invention of television, the phone, cinema, radio, the car, the bicycle, printing, the wheel and so on, but you would think we would learn the way these things work, which is this:
>
> 1. everything that's already in the world when you're born is just normal;
>
> 2. anything that gets invented between then and before you turn thirty is incredibly exciting and creative and with any luck you can make a career out of it;
>
> 3. anything that gets invented after you're thirty is against the natural order of things and the beginning of the end of civilization as we know it until it's been around for about ten years when it gradually turns out to be alright really.[24]

"Ideas are having sex with each other as never before," science journalist Matt Ridley writes in The Wall Street Journal and his book *The Rational Optimist*. He argues that the leap forward humans took over other species forty-five thousand years ago did not come from some "big bang of human consciousness" inside our heads. Instead, he maintains, it came from our public interaction, our invention of tools, and our trade in them, which led to the development of a "collective intelligence."[25] Ridley contends that "all the ingredients of human success—tool making, big brains, culture, fire, even language—seem to have been in place half a million years before and nothing happened." Neanderthals had brains bigger than ours, yet they did not farm or make economic progress. Nomadic people did not settle and advance. "Then suddenly—bang!—culture exploded, starting in Africa," he says. It turns out that "what determines the inventiveness and rate of cultural change of a population is the amount of interaction between individuals." That is, our publicness and our connections bring progress.

The market was the catalyst of civilization, as it caused us to build cities, travel, mix, and interact. "Trade is to culture as sex is to biology," Ridley says. "Exchange makes change collective and cumulative. It becomes possible to draw upon inventions made throughout society, not just in your neighborhood." Ridley sees great hope for our future in the explosion of interaction and the endless, serendipitous sharing of ideas the internet allows. His favorite illustration is an invention that resulted from a conversation between a gastroenterologist and a guided-missile designer: the camera pill. (That would give Alan Westin agita.) Publicness leads to interaction, which leads to innovation. If the interaction of the marketplace and the city got us this far, one wonders where the explosive interaction of the internet will take us.

The Making of the Modern Public

Our concepts of public and private today—and what constitutes the public sphere—are relatively recent, at least in the West. In the Roman Republic, the state belonged to its citizens (a limited constituency, to be sure).[26] Prior to the early modern period—the sixteenth and seventeenth centuries—"public" was synonymous with the state,[27] and the state was synonymous with the king. As Louis XIV liked to say, *"L'état, c'est moi."* The state—that's me. Only men of official stature were public. This construct explains why, to Americans' confusion, what in the United States are called private schools for the privileged are in England called public schools, as they were operated for the children of public men. It is also why in the army, Jürgen Habermas says, the "common soldier—the ordinary man without rank" is a private.[28]

"Privacy is a modern invention," Lawrence Friedman says in *Guarding Life's Dark Secrets*. "Medieval people had no concept of privacy. They also had no actual privacy. Nobody was ever alone. No ordinary person had private space. Houses were tiny and crowded. Everyone was embedded in a face-to-face community. Privacy, as idea and reality, is the creation of modern bourgeois society."[29] Richard Sennett in *The Fall of Public Man* says that the first recorded uses of "public" in English in 1470 associated the word with "the common good in society." Not until seventy years later did it also mean "open to observation."[30] "Privacy" did not enter the German language until the middle of the sixteenth century.[31]

Privacy was not an enviable state. "The word privacy," Patricia Meyer Spacks explains in *Privacy: Concealing the Eighteenth-Century Self*, "derives from a Latin word meaning deprived; deprived of public office; in other words, cut off from the full and appropriate functioning of a man."[32] A nobody, in short. Or, as Hannah Arendt puts it in *The Human Condition*, "A man who lived only a private life, who like the slave was not permitted to enter the public realm, or like the barbarian had not chosen to establish such a realm, was not fully human."[33] Michael Warner, author of *Publics and Counterpublics*, notes the prurient root of privacy: "A child's earliest education in shame, deportment, and cleaning is an initiation into the prevailing meaning of public and private, as when he or she locates his or her 'privates' or is trained to visit the 'privy.'"[34]

Privacy was not assumed to be a good, Spacks says. Indeed, privacy was thought to present a danger to the social order and to vulnerable people—women and children especially—who were better protected from abuse in public view. "What our forebears considered a danger," she says, "we assume as our due." Privacy also separated people from an early ideal of communal responsibility.[35] In 1516, Sir Thomas More argued in his novel *Utopia* that the idyllic society is the transparent society: "With the eyes of everyone upon them, [people] have no choice but to do their customary work or to enjoy pastimes which are not dishonorable." [36] In More's time, everyone worked under the gaze of everyone else. Public business was conducted out of private homes: the cobbler made his shoes there, the alehouse was a house. Privacy in the modern sense was not expected even among the rich, who were surrounded by servants in homes whose layouts required residents to walk through rooms to get to other rooms. The "revolutionary invention of backstairs" in the late seventeenth century separated servants from employers, according to the historian Mark Girouard. [37] The use of interior halls in the eighteenth century enabled rooms to be closed to traffic. [38] The closet, created as a place to lock up one's stuff, later became a retreat for what we would call privacy. In the nineteenth century, Richard Sennett says, Londoners joined clubs not to socialize, as we do today, but to sit in silence, away from the hustle of the city. [39] At the same time, Daniel J. Solove points out, work shifted from farms to factories and offices, turning the home at last into a family's retreat. [40] Thus England, says Philippe Ariès, became "the birthplace of privacy." [41]

Just as privacy was emerging in early modern Europe, so did publicness begin to manifest itself. That is the view of the Making Publics project, a fascinating five-year study by a group of Canadian and American academics. [42] They maintain that new tools—the printing press, of course, as well as the theater stage, art, printed music, printed sermons, maps, and markets—allowed people to collect around interests and ideas, separate from their families, social castes, and work. They were making new publics. McGill University Shakespeare scholar Paul Yachnin, who heads the project, explains that when three thousand people watched *Richard II* in the Globe Theatre and started thinking about what a community should do when its ruler is incompetent, "that seems to be creating the conditions for public participation." They don't need to meet or know one another to have an idea link them together as a public. At first, this coming together did not threaten the powerful. "None of the people in authority think that it matters," Yachnin explains. "It's not a bunch of people speaking truth to power. It's a bunch of people fooling around. That's the reason the authorities let it exist. It never occurs to them that it's changing the shape of their society." Similarly today, the internet's tools and their output were at first dismissed by the power elite—until their impact on media, industries, and governance sneaked up from behind.

In the introduction to the scholarly companion to the project, *Making Publics in Early Modern Europe,* Yachnin and his coauthor, Bronwen Wilson, note that "the boundary between producers and consumers tended to be highly porous in early modernity. . . . Shakespeare, for example, must have started out as a playgoer before he became a player and dramatist." [43] Then as now, the line between creator and consumer was fine. In centuries in between, media came to be the product of corporations, raising the barrier to entry and making the ability to create publics the province of already powerful institutions, not individuals.

Tools can redefine communities. In Reformation Europe, painters lost work when Protestant churches became spare and unadorned, so they turned to making portraits. That gave a public face to their subjects and their communities and allowed outsiders to compare themselves and establish their own identities. Thus the citizens of Venice could recognize the odd Dutch trader visiting town. "Public making is not simply a result of the dissemination of *Hamlet,* engravings, globes, or melancholy," Wilson and Yachnin write, "it is their movement and translation into different media and into diverse places." That movement spawns multiple perspectives. On the internet today, we call this remixing. These are the synaptic arcs in the collective consciousness Ridley writes about—the fruit of public connections.

When people make new publics, they also redefine the idea of the public. Now the king is not the country; The Times is not public opinion; the party is not the body politic; Hollywood is not the culture. A CBC Radio series about Yachnin's project, *The Origins of the Modern Public,* explains that new discoveries—of the Americas, for example— unsettled people's assumptions about their world. [44] Thus even the map and the globe became tools to form and reform publics. Think about how the discovery of the eighth continent—the internet—is changing our perspective of our world today. [45]

The similarities between the impact of change in the early modern period and today are striking. Then as now, new tools empower new actors to create in public and thus make publics. Institutions that held power—the church and the king then; media, corporations, and governments today—find themselves disrupted by their own constituents. Our tools may bear diminutive names—Google, Blogger, Twitter, YouTube, Flickr—but they play no less havoc in the culture than the press, the portrait, the printed song, the explorer's ship, the suddenly round globe, and new marketplaces did in their day. In both the Renaissance and the internet age we hold the promise of breaking out of prior limits, "whether intellectual or territorial," Leah Marcus writes in her essay, "Cyberspace Renaissance." Only today, we explore "without conquistadors."[46]

The Public Sphere

German philosopher Jürgen Habermas writes in an academic style as dense and often indigestible as cold bratwurst.[47] His scholarly followers and critics are often not quick reads either. Yet all who cite him agree that it is Habermas who defines the public sphere and establishes the terms of the debate about its formation and its fate. His framing is useful in discussing where publicness, the public sphere, media, and democracy may go next, after the internet. I'll spare you the sausage making.

In 1962, Habermas published the German edition of *The Structural Transformation of the Public Sphere: An Inquiry into a Category of Bourgeois Society*. It was translated into English much later, in 1989.[48] In it, he argues that it was not until the eighteenth century that public opinion gathered to become a counterweight to the state. Thus civil society came into its own, separate from government. Public opinion coalesced through what Habermas calls "rational-critical debate" among private citizens in the coffeehouses and salons of England and Europe.[49] Britain's first coffeehouse opened in Oxford in 1650, says Brian Cowan in *The Social Life of Coffee*;[50] by the start of the eighteenth century, London alone had three thousand.[51] Known as "penny universities," they became "centers for news culture."[52]

The state was less than receptive to its new critics. In 1784, Frederick II, the king of Prussia, said that "a private person had no right to pass public and perhaps even disapproving judgment" on the actions of government. But in 1792, public opinion was first cited in debate in Parliament. And France's constitutions of 1791 (two years after the start of the French Revolution) and 1793 tied the notions of free speech and publics: "The free communication of ideas and opinions is one of the most precious rights of man," said the first. The second added the right to assemble.[53] The public sphere was indeed coming into its own.

Habermas' critics say he holds an idealized view of the eighteenth-century public sphere. He starts his treatise by arguing that "we call events and occasions 'public' when they are open to all."[54] Yet the bourgeois coffee klatches he admires—and his definition of them as the public sphere—were far from inclusive. They left out women and most everyone but the gentry. "The view that women were excluded from the public sphere turns out to be ideological; it rests on a class- and gender-based notion of publicness," Nancy Fraser says.[55] Michael Warner contends that publics are often formed out of struggle against discrimination, unfairness, and injustice.[56] Thus publics are more likely to start with the people who are excluded from power and property—and from Habermas' twee salons and coffeehouses.

It's debatable whether the discussions in those coffeehouses were necessarily so rational and critical. Was there once a bright, shining period—decades long—when political discussion floated above "the dictates of life's necessities,"[57] unencumbered by emotion and self-interest? I doubt it. Perhaps Habermas is indulging in another form of exclusion: only discussion based on reason and scholarship qualifies as "rational-critical" in his judgment; the rest, he'd sooner ignore. Still, Habermas sets an ideal for gathering political consensus through intelligent discussion of the public good.

Yet as soon as he defines his exemplar of the public sphere, Habermas chronicles its demise, arguing in his masterwork that the public sphere was soon in a state of decomposition. Mass media were the key culprit. Doing the bidding of society's power structure, media tried to influence and corrupt public opinion rather than listen to it. "The world fashioned by the mass media," Habermas says, "is a public sphere in appearance only." He laments that the state took over many functions of society and the family—education, housing, protection. Citizens became clients of welfare entitlements. The family shifted from a productive institution into a mere "consumer of income and leisure time." Thus the separation of state and society melted away.[58] He calls this empowerment of state over citizen "refeudalization."[59] Habermas also complains about state intervention in favor of corporations—engaging in protectionism over free trade and allowing "oligopolistic mergers."[60] He says that opaque markets gave too much power to corporations over consumers while industrial capitalism gave too much power to corporations over workers.

One might think that Habermas would welcome the entry of the internet to break apart mass media's hegemony, give citizens new voice, energize public discussion, and disrupt tyrants and oligopolies. In the rare public comment on the internet—which comes in a footnote to a 2006 talk—Habermas applauds the net's impact on dictatorships but then pays little heed to the value of the conversation there. "The internet has certainly reactivated the grassroots of an egalitarian public of writers and readers," he says, giving the net credit for that much. But then he contends that the web can claim "unequivocal democratic merits" only for its ability to "undermine the censorship of authoritarian regimes that try to control and repress public opinion." That is as far as he is willing to go. "In the context of liberal regimes," he continues (my emphasis),

> the rise of *millions of fragmented chat rooms across the world tend instead to lead to the fragmentation of large but politically focused mass audiences into a huge number of isolated issue publics.* Within established national public spheres, the online debates of web users only promote political communication, when news groups crystallize around the focal points of the quality press, for example, national newspapers and political magazines.[61]

That puzzles me. The man who celebrates the public seems rather dismissive of its members. Habermas venerates the chat in a thousand coffeehouses but denigrates it in a million chat rooms. Fragmentation? That is the complaint one hears most often from mass media that see their audiences migrating to many new competitors. Habermas disdained mass media, but now he longs for quality press as a focal point for discussion. Habermas lamented discussion administered by media, but now he wants quality media to mediate what the public has to say. Most tellingly, he laments the migration of the mass to "isolated issue publics." Yet isn't that dispersal of interests happening because people can now address what matters to them rather than what editors, politicians, or academics think should matter to them?

Meet the real public sphere. We're not neat and orderly or lofty and literary. We are many publics, and now you can hear our diverse and multiple voices, more than could ever be heard in salons or through printing presses. It's cacophonous. Freedom is. It's uncontrolled. That's the point. I believe we are at the next step in a progression of our conception of the public. Today there is the opportunity to break free of the notion of society as a single collection, a mass. As sociologist Raymond Williams says, "There are in fact no masses; there are only ways of seeing people as masses." [62] In *The Sociological Imagination*, C. W. Mills defines the differences between "public" and "mass": [63]

Mass	Public
"Far fewer people express opinions than receive them"; they "receive impressions from the mass media."	"Virtually as many people express opinions as receive them."
"It is difficult or impossible for the individual to answer back."	"There is a chance immediately and effectively to answer back any opinion expressed in public."
"The realization of opinion in action is controlled by authorities."	Opinion "readily finds an outlet in effective action, even against—if necessary—the prevailing system of authority."
"The mass has no autonomy from institutions."	"Authoritative institutions do not penetrate the public, which is thus more or less autonomous."

In Mills' definition of the mass, I see media's worldview. In his definition of the public—written in 1959—I see the promise of the internet and its many publics.

When the Making Publics team first met at McGill University in 2003, one of its members told Yachnin that the assembled academics could not agree with the premise of the project. There could not have been publics in the sixteenth and seventeenth centuries, they said, because Habermas had satisfied them that the public sphere did not emerge until the eighteenth century. After much debate and research, the team came to see that there were publics and the means to make them before there was the public sphere as defined by Habermas. I think they describe a more natural state of society than Habermas does, a state to which we are now returning, of many protean publics. This progression is playing itself out in, of all places, Facebook, as we try to distinguish between the public sphere (*the* public) and the making of publics (*our* publics). We are redefining our idea of "public."

The Public Press

Gutenberg's—and God's—Gift

The problem with media, for most people, has been that somebody else owns them. The media decide where attention should be paid: what news to cover, who gets into print, what gets onto the air. In the old sense of the word, media's proprietors are the public people—they are the privileged. The rest of us, the private people, are those deprived of notice in the media. I begin to see what Habermas means when he laments the re-feudalization of the public sphere: Publicness, once freed from the grip of the state in favor of the people, was soon taken over by another oligarchy, the media. But today, with our new tools for making publics, the people can again become proprietors of media (see: blogs) and the public sphere (see: revolutions in the Middle East). We all have our Gutenberg presses and the privileges they accord. That is what makes it worthwhile to study the development of Gutenberg's era for the lessons it teaches us about the changes in our own.

It would be hard to find a better guide to Gutenberg's transformative power than Elizabeth Eisenstein's monumental *The Printing Press as an Agent of Change,* a two-volume, eight-hundred-page labor of fifteen years' research published in 1979. The book was written before Habermas' masterwork was translated into English in 1989 and before the web browser brought the internet to the broad public in 1994. So we have today an advantage over Eisenstein in being able to spot the parallels between the ages. I called Eisenstein to get an update and ask how she compares the era she studied with that of the internet. She demurred, claiming she's not au courant with the net. She watches her grandchildren sit on her couch when they visit, laptops on laps, talking with friends on Facebook, and says she's not sure what to make of this new society they're creating. The professor is too modest. Eisenstein is in a unique position to inform our grasp of the internet age and its change because she foreshadows it so well in her study of five centuries ago.

Eisenstein contends that the change Gutenberg brought was revolutionary, not merely evolutionary. Printing unleashed "the main forces that have shaped the modern mind."[1] One scholar who agrees with her is Myron Gilmore, who argued in 1963, "The invention and development of printing with movable type brought about the most radical transformation in the conditions of intellectual life in the history of western civilization. . . . Its effects were sooner or later felt in every department of human activity."[2] And, of course, there is Marshall McLuhan. "The difference between the man of print and the man of scribal culture is nearly as great as that between the non-literate and the literate," he wrote in 1962. "The components of Gutenberg technology were not new. But when brought together in the fifteenth century there was an acceleration of social and personal action."[3] It was McLuhan's *The Gutenberg Galaxy* that inspired Eisenstein. "It blew my mind," she recalls. Though she ended up disagreeing with McLuhan's views about technological inevitability, he convinced her of Gutenberg's impact.

Eisenstein chronicles the impact of movable type mostly in Europe, not its earlier introduction in the Far East, where the press did not—as we would say today—go viral. John Man, in his entertaining biography, *The Gutenberg Revolution,* provides a survey of earlier art. Reproducing a font's characters ad infinitum is "an idea so obvious that it occurred to human beings remarkably early."[4] The Phaistos Disc, made about 1700 B.C., has 241 images impressed onto its clay with metal stamps (they remain undeciphered). Ancient Egyptians used wooden blocks to set hieroglyphics onto tile. A key ingredient in printing—paper—was invented in China by A.D. 105 (or two hundred years earlier, according to some accounts[5]) and brought to Korea and Japan five centuries later. The idea of impressing images on paper with stamps seems to have been born in the fifth century, Man reports. In the eighth century, books were being printed from blocks of wood or stone in China, Japan, and

Korea. In 1234, Korea took the lead with the first use of movable type. But the writing systems of all three languages were too complex; calligraphy was still more efficient than printing. With the Latin alphabet and Gutenberg's innovations, printing at last became sustainable and scalable. "One year," says Man, "it took a month or two to produce a single copy of a book; the next, you could have 500 copies a week."[6] This revolution wasn't just cultural, it was economic. In the early modern period, Paul Yachnin's Making Publics project says, culture found customers and a market to support itself, replacing the resources and control of the church and powerful patrons. Now it also found the technology to help it grow.

What's striking about Eisenstein's portrayal of the early days of European publishing is how protean and plastic it was and how its impact could have turned this way or that. In Eisenstein's narrative, printing's influence took time to form, even though printing presses' output was quickly astounding. In the first fifty years after Gutenberg, according to Lucien Febvre and Henri-Jean Martin's *The Coming of the Book,* 20 million books were published[7]—more, by some reckoning, than all the books copied by all the scribes of Europe during the millennium before. Still, Eisenstein says, "for at least fifty years after the shift there is no striking evidence of cultural change; one must wait a full century after Gutenberg before the outlines of new world pictures begin to emerge into view."[8] The changes brought on by the internet today already appear huge in the mirror, but we are still early in this revolution. We ain't seen nothin' yet.

Today, publishers as a breed have so far tried little more than reproducing their old content and business models in new forms, from CD-ROMs to the web to iPads. It was the same in the Renaissance. The earliest publishers made books to mimic the work of scribes, even designing their typefaces to look like scribes' handwriting. Printing was promoted as automated writing. "They appear not to have perceived the printed book as a fundamentally different form, but rather as a manuscript book that could be produced with greater speed and convenience," Leah Marcus says in "Cyberspace Renaissance."[9] They didn't yet see the possibilities.

Early books brought with them the many flaws of scribes' manu-
scripts. At their birth, printed books were not the sacred temples and acts
of perfection on pedestals we see them as today. The first printed books
were filled with errors. "An age-old process of corruption was aggravated
and accelerated after print," Eisenstein says. Printed mistakes could
spread broadly and quickly, but they also could be caught and corrected
faster. Today, the internet is given blame and credit in the same way.
Because of their broader distribution in books, errors became more dangerous. That's why printers were fined for publishing the "wicked Bible"
of 1631 (it omitted the "not" from the Seventh Commandment . . . look
it up).[10] Some contemporaries thought all of Gutenberg's Bibles were
wicked. When his erstwhile partner and funder, Johann Fust, traveled to
Paris to sell Bibles, the book trade called the police on him because "such
a store of valuable books could be in one man's possession through the
help of the devil himself."[11]

Often publishers and authors turned the process of correcting mistakes
through errata sheets and new editions to their advantage. Sixteenth-
century editors, Eisenstein says, "created vast networks of correspon-
dents, solicited criticism of each edition, sometimes publicly promising
to mention the names of readers who sent in new information or who
spotted the errors which would be weeded out."[12] Today, we would call
that crowdsourcing. The mapmaker Abraham Ortelius "received helpful
suggestions from far and wide, and cartographers stumbled over themselves to send him their latest maps," says Lloyd A. Brown in *The Story
of Maps*.[13] Ortelius' community—the public that gathered around his
work—also produced treatises on topography and local history. "By the
simple expedient of being honest with his readers and inviting criticism
and suggestions," says Gerald Strauss, Ortelius made his book "a sort of
cooperative enterprise on an international basis."[14] Publicness opened the
door to collaboration in this, the earliest Wikipedia or OpenStreetMap.[15]

The revered and closed body of ancient thought that had been passed
down from scribe to scribe, generation to generation, mouth to mouth
"was replaced by an open-ended investigatory process pressing against
every advancing frontier," Eisenstein says. The aim of books before had

been to preserve old knowledge. Now books made it possible to gather, compare, analyze, and spread new information. James A. Dewar and Peng Hwa Ang in *Agent of Change,* a book of essays on Eisenstein, say those attributes of books made the scientific revolution possible. For science, Robert Merton argues, is *public* knowledge. Eisenstein says early publishers "encouraged readers to launch their own research projects and field trips. . . . Thus *a knowledge explosion was set off.*" (My emphasis.) The number of known plants, for example, multiplied from six hundred to six thousand.[16] Readers sent seeds and specimens to authors.[17] The publishing process changed fiction as well. After bringing out *Pamela, or Virtue Rewarded* in 1740, its author, Samuel Richardson, revised his work in subsequent editions in response to criticism and advice from a reading group of women he organized.[18] Remember how the creator of *Heroes* did the same on TV, because the net gave him the means to listen to his audience.

The flexible attitude of early publishers alters the way we look at books and media: not as sculpture cut from rock but as still-wet clay. The problem we've had in recent history—from the industrial age on, and not just in publishing but in most every industry and activity—is that we have made mistakes too expensive to admit. That cuts us off from correction and collaboration with our public and from the open increase of knowledge Eisenstein talks about. But the internet—always wet—begins to fix that.

This cultural outlook of openness in printing's early days could just as easily have gone the other way. The explosion of the printed word—and the lack of control over it—disturbed the elite, including Catholic theologian Desiderius Erasmus. "To what corner of the world do they not fly, these swarms of new books?" he complained. "[T]he very multitude of them is hurtful to scholarship, because it creates a glut and even in good things satiety is most harmful." He feared, according to Eisenstein, that the minds of men "flighty and curious of anything new" would be distracted from "the study of old authors." After the English Civil War, Richard Atkyns, an early writer on printing, longed for the days of royal control over presses. Printers, he lamented, had "filled the Kingdom with so many Books, and the Brains of the People with so many contrary Opinions, that these Paper-pellets become as dangerous as Bullets."[19] In the early modern period a few "humanists called for a system of censorship, never implemented, to guarantee that only high-quality editions be printed," Ann Blair writes in *Agent of Change*.[20] Often today I hear publishers, editors, and academics long for a way to ensure standards of quality on the internet, as if it were a medium like theirs rather than a public space for open conversation.

I don't mean to suggest that early books were all changeable and unstable. As Eisenstein points out, the advantage of printing was that it took knowledge that had been diffuse and easily lost in a few handmade copies and made it permanent, consistent, and accessible to many. It was printing, she says, that enabled Thomas Jefferson to collect all the laws of Virginia. "It seems in character for Jefferson to stress the democratizing aspect of the preservative powers of print which secured precious documents *not by putting them under lock and key* but by removing them from chests and vaults and *duplicating them for all to see*."[21] (My emphases.)

Publishing itself moved into the community, leaving monastery walls for loud and noisy workshops where authors had to hang out as their type was set and corrections made. The shops "served as gathering places for scholars, artists, and literati; as sanctuaries for foreign translators, émigrés and refugees; as institutions of advanced learning, and as focal points for every kind of cultural and intellectual interchange."[22] The hodgepodge of people there led to "cross-fertilization of all kinds" and "changed relationships between men of learning as well as between systems of ideas." This evolution, Eisenstein says, was "social as well as intellectual."[23] The shops, like our web today, made new connections.

Print shops were also businesses. "Printing, a ditto device . . . provided the first uniformly repeatable 'commodity,' the first assembly-line mass production," McLuhan says.[24] "The printer and the bookseller worked above all and from the beginning for profit," say Febvre and Martin.[25] Sadly, that did not end up being the case for Gutenberg himself. He went bankrupt, and Johann Fust took over most of his presses and trade.[26] Still, Gutenberg was the protoentrepreneur in what may have been the earliest technology start-up. That is why I have my entrepreneurial journalism students study Man's biography. He takes us through fascinating detail on the science and inspiration required to invent the handheld mold used to manufacture type at scale; the finer points of metallurgy; what it took to engineer paper that could take impressions on both sides; and the chemistry of ink.

Man also explores the capital needs of the business and the publishers' market-driven imperative to find the first best sellers, create cash flow, and pay back investors. In the world of the scribe, says Andrew Pettegree in *The Book in the Renaissance,* supply and demand were aligned: one book, one scribe, one customer. Not so with printing, which required considerable up-front investment for equipment, paper, and labor, with no guarantee of buyers. In the early days printers flooded the market with classical texts, which failed.[27] They tried to shoehorn old media and models into the new form—a pattern we see today. By the third decade of printing, Pettegree says, the newness of printing had worn off. To survive, printers had to create a profitable model based on a new understanding of demand and marketing. To make money, they produced broadsides, pamphlets, how-to manuals, and church indulgences. "The art of puffery, the writing of blurbs, and other familiar promotional devices were also exploited by early printers," Eisenstein says.[28] Books became promotional vehicles, too, for new authors who could now enter the market. Fifty years after Gutenberg's invention, says Pettegree, the printed book as a medium "finally broke free of its roots in the manuscript world."[29]

Print changed how people understood and interacted with their world in many ways. The Making Publics project argues that the mere discovery of a new continent and the maps that documented it altered people's sense of their own environment. "The invention of printing did away with anonymity, fostering ideas of literary fame and the habit of considering intellectual effort as private property," McLuhan says. He adds that the portable book allowed people to read in privacy and isolation[30]—which, Marcus contends, separated the author from the audience, making media impersonal. "As the Gutenberg typography filled the world the human voice closed down," McLuhan argues.[31] With print, the author lost his obscurity and gained authority, while the reader retreated to anonymity and silence.

Books made learning more efficient. Scholars no longer had to travel from town to town consulting rare books to gain knowledge. Now the books themselves could travel and cross-reference one another, making contradictions in thinking easier to spot. This portability of thought "favored new combinations of old ideas at first and then, later on, the creation of entirely new systems of thought," Eisenstein says.[32] Students no longer learned only at the feet of their masters—ending priests' monopoly on teaching—but also on their own from "silent instructors," as books became known. Isaac Newton taught himself math from books.[33] Learning became the province not just of monks but of children.[34]

Books also changed how we remembered. "Rhyme and cadence were no longer required to preserve certain formulas and recipes," Eisenstein says. "The nature of the collective memory was transformed."[35] Books became our database, our memory as individuals and as society. "It is as if mankind had suddenly obtained a trustworthy memory instead of one that was fickle and deceitful," says George Sarton.[36] Today, in the same way, Google has become our personal memory and our universal library.

The press quickly made an impact on the political structure of society. According to Albert Kapr's definitive biography, *Johann Gutenberg: The Man and his Invention,* among the earliest nonreligious publications produced in the great man's shop by his successors—Johann Fust and his son-in-law Peter Schöffer—were political pamphlets. A series of broadsides from each side of a church fight to control the city of Mainz were published on the same presses in 1461, demonstrating from the start that this tool of publicness, like most to follow, was neutral and agnostic. "All these pamphlets were aimed at gaining public support for the respective protagonists and defaming their opponents," Kapr writes. "To the *matériel* of warfare—halberds, rapiers, swords, harquebuses and cannon—psychological weapons had been added, which could be delivered by means of the printing press." Here we see publishing's nascent role in the birth of media, propaganda and the public sphere they would influence.[37]

The printing of maps with set boundaries and the standardization of languages in print began to give shape to national identities. "It is no accident," says Eisenstein, "that nationalism and mass literacy have developed together."[38] Martin Luther's impact on the world came not with his pounding the nail into the door of the Wittenberg church but via the press, as his thirty publications sold more than 300,000 copies between 1517 and 1520.[39] "Luther, himself, described printing as 'God's highest and extremest act of grace,'" says Eisenstein.[40] On this much, Luther and his adversary, Pope Leo X, agreed. Printing, the pontiff said in a 1515 censorship decree, "had come down from the heavens as a gift of God."[41] Or his curse. Luther's Reformation was to be the first revolutionary movement spawned and spread by the printing press as a tool of publicness.

The Making—and Unmaking—of Mass Media

Let us leap forward from Gutenberg and books to the mechanized press and the newspapers it could produce. Now "the press" takes on new meaning not as a machine but as an industry and a public institution. From that moment on, competing forces have battled for control of the press and the ability to speak to and for the public. Newspapers were, at the start of the nineteenth century, organs of political parties and their interests until advertising support allowed them to become economically independent of political ownership. Journalists then began to fancy themselves as autonomous representatives of the public, the intermediary between the people and the state. That is when—*pffffft!*—Habermas' ideal of the civil public sphere deflated. The people could not speak for themselves. They were spoken to.

There's another way to look at this relationship of press and public, James Carey's way. In a collection of essays by and about the Columbia University journalism professor, NYU's Jay Rosen writes that in Carey's view, "The press does not 'inform' the public. It is 'the public' that ought to inform the press. The true subject matter of journalism is the conversation the public is having with itself."[42]

I met Carey once before he died. At the time, I was new to the journo-academic circuit of conferences and panels. I came off the dais at a university blatherfest, having badgered the traditional journalists in the room about the idea the internet had introduced—or so I thought—of news as a conversation. I sat down next to Professor Carey, and he whispered to me that he'd based his career on the idea of the journalistic conversation. Carey had long taught that the American Bill of Rights is "an injunction as to how we might live together as a people, peacefully and argumentatively but civilly and progressively."[43] The Founding Fathers, Carey said, charged us to "create a conversational society. . . . Other words might do: a society of argument, disputation, or debate, for example. But I believe we must begin with the primacy of conversation. It implies social arrangements less hierarchical and more egalitarian than its alternatives."[44] Carey, like Habermas, sought the civil ideal of rational, critical debate as the central nervous system of democratic society. He expected the press to reflect and not just enlighten that public conversation.

"Public," Carey said, is "the God term of the press, the term without which the press does not make any sense." So the press says that it guards the public good, reports public opinion, educates the public, serves the public's right to know. But Carey, like Habermas, saw a corruption in media's use of the public. Carey's devil was the public-opinion industry. Polling "was an attempt to simulate public opinion in order to prevent an authentic public opinion from forming." Polls reduce our perspectives to numbers to be manipulated, robbing us of the nuance and complexity of conversation. "A press independent of the conversation of culture," Carey said, is likely to be "a menace to public life and effective politics."[45] Strong words but true. As a journalist, I was not taught that my job was to foster, collect, and disseminate the public conversation. I was told that my role was to inform the public, which implied that the public was uninformed. We journalists raised ourselves up over the public and separated ourselves from it, believing we were objective, opinionless people, purer in that respect than the politicians we covered or the citizens we served.

Anthropologist Jack Goody argues that human history is not so much a story of the struggle over the means of production but instead over the means and modes of communication.[46] That certainly can be said of the twentieth century, when the fight to control communication to the public sometimes took on the air of a holy war. When newfangled radio threatened newspaper publishers, print barons resorted to "the invocation of sacred rhetoric," says Gwyneth L. Jackaway in *Media at War*. "Radio journalism, they warned, posed a threat to the journalistic ideals of objectivity, the social ideals of public service, the capitalist ideals of property rights, and the political ideas of democracy. . . . Thus, as a means of defending their own interests they invoked the interests of the nation."[47]

In the early days of radio, newspapers treated the medium as a curiosity, giving it coverage and attention. Then publishers realized that radio could be trouble. "What had started as an enjoyable pastime for young boys in their basements was becoming a medium to reckon with," Jackaway writes. Publishers complained that radio news departments didn't have enough reporters and editors to uphold the high journalistic standards of print. They complained that radio was cutting into their sales. They complained that radio was violating their copyrights.

The same story played out when television threatened newspapers. Editors called TV reporters "parasites" and tried to keep them out of the White House pressroom.[48] In the 1980s, newspapers went after phone companies when they got into the content business, pre-web, offering information services via phone calls.[49] Today, the pattern holds once more as old-media titans complain about new-media upstarts, dismissing bloggers as opinionated amateurs who don't uphold the same standards as the industry. They worry, like Erasmus, about a glut of content distracting the people who had been their devoted audience. In many a debate over the future of media industries, I have heard executives from legacy companies remind us that so far no new medium has killed an old one: after newspapers, we still have books; after TV, radio survives. But today, newspapers are expensive, capital-intensive, now-uncompetitive enterprises; they could die. Magazines are faltering. Radio, at least in the United States, is a shadow of its former self. In 2010, former Massachusetts Institute of Technology visionary Nicholas Negroponte gave the printed book five more years.[50] I hope he's wrong, not for me but for my daughter, Julia. Since she was little, she has wanted to write books. I hope they are still around for her to write. But to hedge her bets, Julia also taught herself video editing on Final Cut.

This shift in technology, some say, could change not only the economy and society but also us and how we think. Author Nicholas Carr infamously asks and answers his own question: Is Google making us stupid?[51] He argues that the net curtails our "deep reading" of books and the "deep thinking" that results. That assumes books are necessarily the only or best way to stimulate thought. Eric Schmidt's response to Carr's question: Aren't we smarter now?[52] I will confess I read fewer books postnet, but I believe I am more curious thanks to the net—because it causes curiosities and because it can so easily satisfy them. I will work through an idea over many blog posts and many weeks or months, using the interplay with my readers—the conversation—to challenge and, I hope, improve my thinking. Writing in books can be deep or shallow—as ideas can be deep or shallow online.

This disagreement over the web's impact on our thinking feeds a discussion about whether technology is changing how our brains operate.[53] As much as I am prone to see the internet as an agent of change in every corner of life, even I don't think it alters biology. I do, however, believe the net changes how we see our world and interact in it—just as the invention of print did—and it is similarly unsettling. "It is both amusing and comforting to recognize how closely our uneasiness with the unleashing of previously fixed text into the nebulous free fall of cyberspace approximates the anxiety experienced by Renaissance authors as they surrendered their writings into what appeared to them as the impersonality and uncontrolled dispersal of print," Marcus says. Today, writing on computers and constantly changing our words "is eroding the distinction between manuscript and printed book, thereby giving our own era special access to a Renaissance state of mind in which the distinction had not yet been clearly established," she writes. Just as early modern readers "needed considerable time to adjust to the decreased auditory stimulation of the printed book," we are surrendering the familiar and comfortable "tactile and visual elements of book reading" to computer and tablet screens or audiobooks. The fact that we can read and write simultaneously, Marcus argues, breaks the line between originals and the commentary or the remixes they inspire. It "seriously diminishes the reverence in which authorship has been held for several centuries." Whether in books or in journalism, Marcus asserts, we see this breakdown of old "hierarchical systems of authority" and hear people wondering what they'll do without it.[54] That, as my friend Jay Rosen says, is the moment when readers become writers and writers become readers.

A group of Danish academics coined a magnificent phrase for this paradigm: the Gutenberg Parenthesis.[55] Before Gutenberg, knowledge was preserved by scribes and passed around orally, remixed along the way. The Gutenberg centuries—1500 to 2000—were an age "dominated and even defined by the cultural significance of print," the Danes say. Authority during the Gutenberg Parenthesis "rested on the mastery of the accumulated canon of wisdom lodged in books (in Bacon's words, books were 'ships of time' bearing precious cargo through the ages)." Whether in science or fiction, our thought patterns, at least in the West, came to mimic linear print: We thought in straight lines. "The line, the continuum—this sentence is a prime example—became the organizing principle of life," McLuhan writes.[56]

Digital content, on the other hand, "is infinitely changeable and flexible," the Danes say. "Recognising a text not as a final product in an edition of a mass-produced printed book, but as a never-stopping ongoing process—blog, wiki, etc.—owing its existence not to a privileged author but to the contributions of very many proximate but unseen hands, will have the greatest consequence for cognition. From the finished product of the book we are on the way to the never-finished, multi-originated, and multi-media shifting work in eternal progress." Authority, they say, will now spring from mastering "the permanence of change."

Before the Parenthesis media were handwritten, oral, shared, affected by the process of distribution, often anonymous, supported by patrons, and emphasized preserving ancient wisdom over gathering new knowledge.

Inside the Parenthesis media were written, linear, fixed, permanent, authored, owned, packaged as products, commercial, with a distinct beginning and a clear end.

After the Parenthesis our experience of media once again becomes conversational, open, shared, remixed, based once more on process more than on product, collaborative, amateur, and endless.

These changes have occurred in the context of a much larger shift from an industrial economy to a digital economy with information as a new currency. That changes much more than media. How should we react to this change? There are only two possible choices: to resist it, which is futile, or to understand it and find the opportunity in it. "There is no point in dwelling on the dark potentialities of the new technology when we have, in fact, no way of predicting its eventual effects," says Marcus. "Renaissances happen by infrequently enough that they should be enjoyed in the process. I, for one, await the Cyberspace Renaissance with great interest, and hope to live to see its zenith."[57] Amen.

What Is Privacy?

How Do You Define Privacy?

That turns out to be a surprisingly difficult question to answer. Try it. Write your definition—if you can come up with one. Then come back later, after reading many others—and all the questions they raise—and see whether yours still holds. I warn you: There's a maze ahead.

My own first attempt to define privacy when I began this project was, I now realize, a cop-out. I wrote on my blog—echoing Mark Zuckerberg—that the issue isn't privacy but control: control of our information and our identities. Sounds neat and tidy. But privacy isn't. What is our own information? Is it what we say, do, like, buy, or make? Is it where we go or whom we know? When our information joins with others', who owns the result: the crowd or each of us? Why is information private: because it comes from behind closed doors, because of the type of information it is, because of the conditions we put on it, or because we fear it getting out? Or is the key factor the use to which others put that information—to serve us, to surveil us, to sell to us, to judge us? Doesn't privacy mean different things in our homes, jobs, and communities? Your privacy is not my privacy. So how can we have one definition?

"Privacy seems to encompass everything, and therefore it appears to be nothing in itself," says Daniel J. Solove. In his book *Understanding Privacy,* Solove quotes others lamenting privacy's "protean" form and its "embarrassment of meanings."[1] "Perhaps the most striking thing about the right to privacy is that nobody seems to have any clear idea what it is," says Judith Jarvis Thomson.[2] "Privacy is a chameleon-like word" used to cover a wide range of interests and "to generate goodwill on behalf of whatever interest is being asserted in its name," says legal scholar Lillian

BeVier.[3] "Privacy has become as nebulous a concept as 'happiness' or 'security,'" writes Raymond Wacks.[4] Solove notes that it's easier to articulate the interests of what he calls the other side of privacy and I call publicness: "free speech, efficient consumer transactions, and security." But I don't intend to pit privacy against publicness—because, again, they are not opposites; they interact with each other. We cannot define privacy as that which isn't public.

I want to define privacy in its own right. Many others have tried before me. Alan F. Westin, in his 1967 book *Privacy and Freedom,* proposes this: "Privacy is the claim of individuals, groups, or institutions to determine for themselves when, how, and to what extent information about them is communicated to others." Thus it is an attempt to restrict what is learned about me. But doesn't that depend on how it is learned? If you see me in a parade and tell others about it, how is that private? Westin also says that privacy is "the voluntary and temporary withdrawal of a person from the general society through physical or psychological means." He acknowledges that privacy reduces participation in one's community and consequently has an impact on publicness. "Thus each individual is continually engaged in a personal adjustment process." That is to say, there isn't a static definition.[5]

In 1960, William L. Prosser, a leading American scholar on tort law, separated privacy into four torts:

1. Intrusion on one's seclusion or solitude or into one's private affairs

2. Public disclosure of embarrassing private facts

3. Publicity that places one in a false light in the public eye

4. Appropriation of one's name and likeness

These four kinds of invasion of four different interests are tied together by the name "privacy" but "otherwise have almost nothing in common" except interference with the right to be let alone, Prosser says.[6]

In his own taxonomy of privacy, Solove separates violations into four activities: "Information collection, information processing, information dissemination, and invasion."[7] He then lists various flavors of privacy: "Limited access to oneself . . . control over personal information . . . secrecy . . . personhood—the protection of one's personality, individuality, and dignity . . . intimacy." It becomes daunting to chart all the variations.

"The right to life," Louis Brandeis and Samuel Warren write in their 1890 essay, "has come to mean the right to enjoy life—the right to be let alone."[8] Their classic framing implies a right to solitude. Privacy here becomes almost hermitic. Violations of privacy, in their view, are oftentimes nuisances that interrupt our quiet. This idea of privacy also implies that once we venture out from that seclusion, control is gone. It's not a satisfying definition.

Because privacy was not specifically protected in the Constitution and its amendments, Brandeis and Warren—like privacy scholars who followed them—had to infer privacy in various clauses of the Bill of Rights and in other law. They had to create a right to privacy. To do so, they call on the Fifth Amendment's protection against self-incrimination and note that common law gives each individual the right of determining "to what extent his thoughts, sentiments, and emotions shall be communicated to others." They read privacy there. Other scholars find elements of a right to privacy in the First Amendment (free speech), the Third (protection from quartering troops in one's home, that is, one's castle), the Fourth (protection from unreasonable search and seizure), and the Ninth (protection of rights not enumerated in the Constitution). Brandeis and Warren also rely on copyright, saying a person "is entitled to decide whether that which is his shall be given to the public." They argue that the doctrine is rooted in "the right to one's personality." In setting a right to privacy, they must also set its limits. They acknowledge that such a right could not prohibit publication "of matter which is of public or general interest." So privacy would benefit people "with whose affairs the community has no legitimate concern." That is, private people. Tough luck, public figures.

Brandeis and Warren also grapple with a keystone question of the privacy debate: harm. What are we guarding against? The real damage, they say, of violating privacy is not physical or even economic but emotional—"injury to feelings." What will other people think of us, and what is our fear of that? This concern says less about us than it does about the other people (or our perception of their perception of us). Privacy is a hall of mirrors. I warned of the maze. In the end, Westin says, Brandeis and Warren's essay "was essentially a protest by spokesmen for patrician values against the rise of the political and cultural values of 'mass society' "—and mass media.[9]

For his part, Westin demarcates zones of privacy. The inner circle contains our "ultimate secrets . . . beyond sharing with anyone." The next circle holds "intimate secrets," which we share with those whom we trust. Next out is our circle of friends. Out at the edge are circles of "casual conversation . . . open to anyone"—that is, the public. That resembles the model of privacy we follow as we try to manage our use of Facebook: what we wouldn't tell anyone, what we'd tell friends, what we'd tell everyone. The worst that can happen, Westin says, is "that someone may penetrate the inner zone and learn his ultimate secrets, either by physical or psychological means. This deliberate penetration . . . would leave him naked to ridicule and shame and put him under the control of those who knew his secrets."[10] So his definition, too, is built on emotional and social fear.

If privacy is about fear—and there's nothing wrong with that, as fear is a means of worst-case analysis, of establishing our apprehension and protecting against it—Brandeis and Warren do not address other fears and other enemies, some more modern than the rest, which I hear in the discussion of privacy:

- Identity theft, leading to loss of money, damaged financial reputation, and considerable hassle. I'd say that's less an issue of privacy than of larceny.

- Unauthorized use of one's image or reputation for others' commercial purposes (rather than news or free speech)—

that is, use of your image, name, or reputation to promote a product. I see this, too, as an issue of theft, but it has been notched into the privacy debate.

- Telling lies about us has also become a branch of the privacy debate. I see that more as a question of defamation, as one can lie about me while knowing nothing of my private life or thoughts. I don't need to know you to baselessly accuse you of beating cats.

- Salesmen and spammers finding and bothering us or spooking us with their knowledge. ("Why are they offering me cheap Viagra . . . Is there something they know?") Though isn't that more a matter of disturbing the peace?

The most prevalent and menacing concern in the privacy debate—especially since the middle of the last century—is fear of government: government surveillance, government invasion, government prosecution, government control of our lives (enter: abortion and 1973's *Roe v. Wade,* with privacy as a factor in government interference in a couple's intimate relations and a woman's right to choose). Though the specter of fascist and communist totalitarian regimes casts dark shadows here, it is Orwell's *1984,* published in 1949, that has become the touchstone and perhaps a catalyst for worry about the state and privacy. The Washington Post editorialized that what was most frightening about Orwell's world was "the complete abolition of privacy." [11] Orwell inspired the kinds of ominous, overstepping predictions that Time magazine has always specialized in: "Televisionaries," Time said in 1950, "confidently forecast the day when every home will have its private network (so mother can keep track of the kids) and telephones will come equipped with TV screens. But there is a chill in the air: in that event, would Big Brother and his thought-controlling telescreens be far behind?" [12] You tell me, citizens of that future: Is he?

It was government's meddling in privacy that provided Brandeis the opportunity to help bring privacy from theory to law after he rose to the Supreme Court. In 1928's *Olmstead v. United States,* the Court's majority ruled that government wiretapping was not an unreasonable search or seizure. Brandeis dissented. He declared that the Constitution "conferred, as against the government, the right to be let alone—the most comprehensive of rights and *the right most valued by civilized men.*"[13] (My emphasis. Do you think the right to privacy or speech is our most valued?) Brandeis left the bench in 1939, but the court incorporated his rationale for privacy in 1967 when it overturned *Olmstead* in *Katz v. United States.* In this post-Nazi, post-Orwell, Cold War case, the court, under Chief Justice Earl Warren, embraced Brandeis' doctrine, declaring that attaching an electronic eavesdropping device to a phone booth constituted unreasonable search and seizure and, thus, an invasion of privacy. In his history of privacy, Frederick Lane says that in the one hundred sixty-six years before Warren became chief justice in 1953, the word "privacy" was used in only eighty-eight high-court opinions. In Warren's court, it appeared in one hundred seven opinions. Since then, it has been cited in more than five hundred thirty-five.[14] It was a long time coming, but we now have a legal right to privacy in the United States. And it is expanding. The question is: How far will that protection reach?

Privacy is more than a legal question of the government versus the individual. Privacy is a key factor in a society's definition of itself: of the relationship of the person to the community, of the customer to the company, of the limits of ownership, of the rights of the individual. No one I found in my research does a more thoughtful job of portraying the breadth of privacy's domain than literary scholar Patricia Mayer Spacks, author of the social history *Privacy: Concealing the Eighteenth-Century Self.* She sees privacy as a matter of choices. "The notion of privacy develops from a simple concept of being left alone into a way of condensing ideas about autonomy and integrity,"[15] she writes. Privacy "raises questions about the proper balance between responsibility to oneself and to other people." Spacks notes that "privacy, whatever its definition, always implies at least temporary separation from the social body." Or, as sociologist Arnold Simmel says, "We live in a continual competition with society over the ownership of our selves." Privacy is a tug-o'-war with the values of a communal society. Because social relations impose moral discipline (nobody wants to wear the scarlet A), privacy can raise questions of secrecy—"What is she hiding?"—and that can imply danger, Spacks says. "The person who claims the right to be alone, or even to keep things to herself, might meditate bad deeds or entertain bad thoughts, and *no one would know*. . . . Desire for privacy might imply selfishness or irresponsibility."

Privacy also creates voyeuristic temptation: the "eagerness to penetrate the privacy of others." It's not the conversation we can hear that interests us as much as the one we can barely hear. Why? Well, we're curious. But Spacks carries the question another step. "The idea of privacy always carries about it some aura of the erotic," she asserts. "Those 'Privacy Please' signs on the outside of a [hotel] door suggest the possibility of couples at play inside the rooms they protect. . . . So the subject of pornography becomes closely implicated with that of privacy." But that's wishful voyeurism. Most times, privacy doesn't protect the scandalous; it more likely guards the mundane and the monotonous. Spacks recognizes privacy's double-edged sword: "If privacy implies freedom *from*—from watchers, judges, gossips, sensation-seekers—it also connotes freedom *to*: to explore possibilities without fear of external censure. Privacy can constitute a form of enablement." [16]

In *The Secret History of Domesticity*, Michael McKeon quotes dramatist John Dennis saying in 1720 that "Nothing was more a Man's own than his Thoughts and Inventions. . . . The Money that is mine, was somebody's else before, and will be hereafter another's. Houses and Lands too are certain to change their Landlords. . . . But my Thoughts are unalterably and unalienably mine, and never can be another's." [17] Well, yes, until the moment when one shares a thought. Once shared, it cannot be taken back. Unlike property, a thought can be held by many at once. If information is created through an interaction with another, Solove asks, which one of those people owns that information? He points out that our personalities are not private but are a public expression of ourselves. Here we see the constant push-and-pull between solitude and interaction in the privacy debate. Legal scholar Richard Parker takes the equation to its extreme in his definition of privacy: "Control over who can see us, hear us, touch us, smell us, and taste us, in sum, control over who can sense us, is the core of privacy." Solove responds correctly that Parker's rule would make any interpersonal contact an invasion.[18]

In much of the current discussion of privacy, one sees hand-wringing over the ability of computers and databases to put together information about us. "One isolated piece of data about an individual is often not very revealing," Solove writes. "Combining many pieces of information, however, begins to paint a portrait of our identities."[19] It becomes harder, then, to define specific information as private if you don't know what it adds up to. That I buy fertilizer isn't telling until I go to a web site about making bombs with it. Your mother's maiden name may not be interesting until I use it to get a hint to your password for your bank account.

Rather than dealing with the specifics of privacy's violations, some scholars try to address the question in a much larger context. Solove says we should concentrate on harm: the problems we are trying to solve and the activities that are disrupted by a violation of privacy.[20] NYU's Helen Nissenbaum suggests a framework built around context. The author of *Privacy in Context,* she argues that we have a "right to appropriate flow of personal information," and she draws complicated matrices of privacy depending on what role we play with whom under what circumstances and norms.[21] In her view, privacy is dependent on a myriad of factors: who tells what to whom how, why, and when; what relationship they have; what expectations there are. A bit of health data has different impact when shared in health care, work, family, and social contexts. Nissenbaum's thinking is good, but trying to put a set of rules against all the possible permutations of information flow looks too complicated to manage and enforce.

Most of these definitions express privacy in negative terms: preventing the realization of fears. There are also many positive reasons for privacy. "Privacy is a special kind of independence," says Clinton Rossitor. It gives one the space and freedom to create and experiment. "The free man," he continues, "is the private man." Westin says privacy also grants people "a chance to lay their masks aside for a rest." [22] But does that not assume that the only time we are authentic is when we are alone and unmasked? If so, that would be a tragic standard for our relationships.

At the same time, privacy grants the power of anonymity. Many believe anonymity ruins discussion on the internet as trolls hide behind pseudonyms to shoot snarkballs at victims. But anonymity also emboldens and protects whistleblowers and revolutionaries. Alan Westin contrasts seventeenth-century English licensing laws, which required publications to bear the names of authors and printers, with the U.S. First Amendment, which protects all speech, including anonymous speech. He quotes historian A. J. Beveridge's account that between two decades at the turn of the nineteenth century, six presidents, fifteen cabinet members, twenty senators, and thirty-four members of Congress published anonymously or under pen names. [23] The anonymity privacy allows is a privilege that can be abused, but it is a necessary right nonetheless.

The worst definition of privacy, the one I also hear quite often at privacy conferences and in conversations, is contained in one word: "creepy." Internet applications or ad tracking or RFID chips are called creepy. Google puts its Street View camera on a bike to take it to places cars can't go, and that's creepy. [24] Almost every discussion of facial recognition software ends with "creepy." [25] In one of his all-too-quotable quotes, Google's Eric Schmidt says the company's "policy on a lot of things is to get right up to the creepy line and not cross it." [26] That itself was called creepy. When I hear "creepy," I've taken to stopping the conversation and asking the person who uses it to define the word, the context, and the harm. Shrugs ensue. "I don't know. I just don't like it. It's . . . it's creepy." It is an emotional response to the unknown, to what could happen. Though we've seen that privacy is often about feelings and fears, emotion alone is not the proper basis for regulation of new technologies and industries and speech.

Do you feel any closer to a definition of privacy? I don't. I see a confused web of worries, changing norms, varying cultural mores, complicated relationships, conflicting motives, vague feelings of danger with sporadic specific evidence of harm, and unclear laws and regulations made all the more complex by context. Perhaps a better way to tackle this question is not to ask what privacy is but instead to ask what we need to protect and how.

How Do We Protect Privacy?

At South by Southwest, the interactive conference in Austin where both Twitter and Foursquare took off, danah boyd (she prefers lower-case) packs the hall as she gives stern admonitions and good advice to adults about teens' privacy: "Each of you—as designers, as marketers, as parents, as users—needs to think through the implications and ethics of your decisions, of what it means to invade someone's privacy, or how your presumptions about someone's publicity may actually affect them. . . . How you handle these challenging issues will affect a generation. Make sure you're creating the future you want to live in." [27]

Conventional wisdom today is, of course, that privacy is dead. The internet wounded it. Facebook killed it. In 1999, Sun CEO Scott McNealy infamously told us, "You have zero privacy anyway. Get over it." [28] But as we've seen, privacy has no shortage of protectors. Privacy is far from dead. I think we could end up with more privacy regulation affecting more of life than ever—and not just in the digital sphere. I also believe we are each more aware of our privacy than we used to be. As a result we're more likely to protect ourselves. "People are actually much more thoughtful about privacy today than they ever were," boyd says, "because they go out of their way to find it." We appreciate privacy because it's harder to keep.

Conventional wisdom also holds that young people don't care about the loss of privacy because they've already abandoned theirs. That, too, is wrong. A 2007 Pew Internet & American Life survey of young people found that among the 55 percent with online profiles at the time, two-thirds limited public access to them; 46 percent sometimes gave false information to protect themselves or to be playful; and 91 percent used social networks to stay in touch with people they already knew—that is, they weren't using social services to interact with strangers. Pew found that many teenagers are savvy about privacy protection. "Many, but not all, teens are aware of the risks of putting information online in a public and durable environment," Pew says. "Many, but certainly not all, teens make thoughtful choices about what to share in what context." [29]

boyd thinks we underestimate the young. But then boyd is an unconventional thinker. For her doctoral dissertation at the University of California at Berkeley, she studied how American teens socialize in networked publics. At Microsoft Research, she studies social media, surveying and talking with teenagers to understand how they interact online. She has written a forthcoming book, *The Social Lives of Networked Teens,* busting myths about adolescents, including notions that social media is addictive; that the internet is a dangerous place; and that young people are digital natives.[30] Youth aren't born with the net in their DNA, she says. They, too, must learn how to live online. Like all of us, they learn how to protect their privacy through experience.

boyd has shown me how teens make sophisticated use of social tools to do what they want to do, often to hide in plain sight. She tells the story of a girl whose mother followed her on Facebook—and the girl was fine with that. But when she broke up with her boyfriend, the girl wanted her friends to know while not alarming her mom and unleashing a flood of maternal concern. The girl decided to post song lyrics on her Facebook Wall, and she chose a happy song, "Always Look on the Bright Side of Life," giving Mom no cause to worry. It appeared happy, after all. But her friends would know that this song was sung in *The Life of Brian* just as the main character was to be executed. That was her signal that something was off and she needed sympathy. She used the tools available to accomplish her goal of controlling not the flow of information but its meaning.

As teens navigate between private and public, boyd wants to make sure they make informed choices. She cautions, for example, that we are seeing "an inversion of defaults when it comes to what's public and what's private." That is, the rules of any given social encounter can no longer be assumed. In real life, when you have a one-on-one conversation in a hallway, the discussion may be private though held in a public place. If others walk up, you can decide whether to include them or change the subject. What you say there won't be spread unless someone in the group does so. Your words would become, as boyd says, "public through effort." On a Facebook Wall, the conversation is instead "public by default, private through effort." A young person holding a conversation there may be fine with that—until it's time to apply to an Ivy League college and worries set in. So a child from the supposedly public generation may need to be more private than you or I. At the same time, boyd says, an adult from the supposedly private generation might just now be learning that "blogging and tweeting open up powerful doors for them." So make no assumptions about young people living in the open and their elders hiding behind doors. We're all experimenting and discovering our limits of privacy and publicness.

boyd sees some solutions to privacy problems as legal, some technical, some social. Her legal argument is unconventional but convincing. "Privacy," she says, "isn't just about controlling the access to information but controlling how it's used, how it's interpreted." The problem is less the gathering of information and more what is done with it. Her example: "If you walk into my office applying for a job, with one quick look I'm going to be able to get a decent sense of your gender, your race, your age." Antidiscrimination law doesn't forbid her from *knowing* these bits of information about me. Instead, it forbids her from *using* them against me in hiring. Of course, she could still deny me the job because of my gray hair. But if she is caught in a pattern of discriminating against applicants on the basis of age, she can be sued.

Regulating the *use* rather than the *gathering* of information is a wise strategy. If we keep chasing around trying to stop information from being revealed or gathered, we'll find ourselves in a perpetual game of whack-a-mole. We'll be telling people, companies, and governments they aren't allowed to know what they have already heard, seen, or read. That is the absurdity U.S. pharmaceutical companies face today. When I have suggested to drug-company executives that they should find new ways to listen to patients, they tell me their lawyers won't allow it. If drug companies hear of problems with medicines and don't react appropriately and immediately, the companies increase their liability. As a result, they avoid conversation. They stick their fingers in their ears. There is the fruit of the law of unintended consequences: These companies won't hear the patients and their problems, needs, and ideas. We're poorer for that.

Now consider insurance companies that may watch the chatter among patients online. "Do they have the right to use that information?" boyd asks. "I think the answer is no, regardless of whether or not they have access to it." The consequence of restricting usage: "If you can't use the information, it makes a lot less sense to try to find ways to access it." boyd thinks college admissions offices should not be allowed to use on-line material and neither should employers—if the material was shared for personal reasons. In Finland, Googling prospective employees is il-legal.[31] In Germany, legislation has been introduced to outlaw the use of information that is too old or over which the employee has no control.[32]

boyd and Germany are bringing another factor into their privacy for-mulas: context. Why, where, and how we share information; with whom; for what reason; with what expectation—all these factors, as Helen Nis-senbaum says, need to be taken into account when the information is used. If you turn me down for a job or insurance or college, you might need to ask yourself why and where you got your information and whether it was intended to be public. Regulation can bring unintended consequences. These days, the law makes employers loath to give honest job references about former employees for fear of volunteering damaging information and getting sued. So they give nothing but name, rank, and serial number. The result can be that employees with problems just get passed around from wrong job to wrong job, failure to failure. Is it right to force employers not to research employees before hiring them?

Context is complex. A teacher is turned down for a job because the principal at a private religious school finds a picture of her partying on Facebook. If it's not illegal to drink and she doesn't do it in front of students—she was sharing this picture with friends—why should that be a problem? Just because the principal disapproves? Hasn't the teacher's privacy been violated? Is the problem here the teacher's behavior, the picture of it, the principal's scrutiny, the principal's policies, or none of the above? What if the teacher makes a lewd gesture at the same time and, because the photo is available for others to see, her young students could also see it? Is the problem the gesture and the picture or that students this age shouldn't be on Facebook snooping on Teacher?

Some hospitals are not just banning smoking at work but are trying to ban smokers from jobs there.[33] Say you're a nurse. A friend posts a picture of you holding a cigarette at a party, and you are fired. Did your friend violate your privacy by posting that picture? Did Facebook violate it by allowing someone to make public a picture of you from a private event? Did the employer, by using information garnered from a private context, now made public, violate your privacy? Or did you simply get caught violating a rule you had agreed to? What if the event were a street party, thus held in public and your boss saw you? Could you argue that you had an expectation of privacy even on a public street? That's the argument I get into with Germans when they contend that one may have an expectation of privacy anywhere. How does anyone else know that is the expectation? And doesn't that allowance risk making everything that's public private? William Prosser wrote in 1960, "On the public street, or in any other public place, the plaintiff has no right to be alone."[34]

We are seeking rules, but in a changing environment, while technologies and our behavior using them are still new, it's difficult to set statutes yet. That won't stop lawyers. Every day when we "click to accept," we agree to rules we don't read and likely wouldn't understand if we did slog through them. In the hands of lawyers, a web cookie begins to sound like a hand grenade. That needn't be so. The principle at work here, in the words of author Cory Doctorow, is informed consent. How can we consent to a site's rules and agree to its procedures if we're not informed? Lawyers don't inform; they obfuscate. When privacy controversies erupt, it does no good to point to forty-six pages of fine print and argue, "It's in there." Facebook learned the hard way that it had to simplify and clarify its privacy policy and tools. It learned that sudden changes in policies only raise suspicion. Every executive in a company, I think, should know its privacy policy and be able to explain it to an elderly aunt or a small child. The wise company would survey users to make sure they under-stand what information is gathered about them, how and why it is used, and what control they have. That is a privacy audit made not from the perspective of compliance with regulations or legal liability but simply to make sure that users are informed. In 2011, in a settlement with the Federal Trade Commission over privacy missteps with its Buzz service, Google agreed to regular independent privacy audits.[35] That's not a bad idea for any organization.

boyd agrees that young people must be equipped to make informed decisions about their own privacy. But she's not saying we need to wrap them up in rules. She argues, in fact, that we overprotect youth. "We've done an amazing job of trying to regulate them, regulate their sexuality, regulate their access to public space, regulate everything that goes on about them and give them no way of coming of age in a healthy man-ner. That worries me." We segregate young people from adults and the public sphere, she says. When she was young and a member of the first generation to grow up online, boyd says, the internet was her saving grace because it let her talk with adults around the world and figure out that world. But we've extended our stranger-danger trepidation to online, with a vengeance.

A product of that worry is the Children's Online Privacy Protection Act (COPPA), the law that puts strict controls on what sites can do with children under thirteen.[36] COPPA requires that if a site wants to gather information from children—name, email address, hometown—it must receive written consent from parents. On its face, that might sound like it makes sense, but boyd points to more unintended consequences. COPPA teaches children to lie—even grandparents get kids to fib, she says, so they can sign up for email or online calling services and stay in touch. On the internet, every child is fourteen. COPPA results in young people being underserved with content and services because companies don't want to be liable. More heartbreaking, boyd says, is that COPPA scares medical organizations that want to help children who are suicidal or have eating disorders from getting "anywhere near" those issues. In boyd's view, we're concentrating on the wrong end of the transaction.

"I don't want more control over the teens themselves," she says. "I want more control over those who hold power over them."

Control. Privacy keeps coming back to that word. How will we achieve it? As boyd says, some solutions are legal, some technical, some social. In an era of change, the three won't be in sync. On a recent trip to Amsterdam, I was astounded when, as we came into the city, my cab driver slowed to the posted speed limit—and not a km/h more—as did everyone else. No one passed anyone. He explained that sensors watch each car. If the machine catches you going over the limit, you'll automatically be ticketed. So all obey. On the one hand, that is enviable: technology results in safer roads—with less need for police and their expense—and greater order. Isn't that what we want for society? But in the U.S., we bridle at the idea of cameras and computers spying and finking on us. In truth, we all speed and cheat a little, and we don't want to get caught. Put another way, to judge by our behavior, most drivers on the road believe that speed limits are set too low. So the limit's the lie. If technology is better able to monitor our adherence to rules, it's not our privacy that's violated, nor is technology the issue. It's that our beliefs and behaviors don't match our laws. Technology only exposes that gap.

So what do we do about that? We wink and tell technology to butt out. We decry sensors and cameras on the road as Big Brother government invading privacy. We say that black boxes in our cars recording our actions would be invasive. When Google announced that it had driven 140,000 unmanned miles in computer-driven cars, I heard many fellow drivers—especially, in my unrepresentative survey, younger drivers—say they would resent and resist the loss of control.[37] (The reaction from orderly Dutch media: "Finally!"[38] The reaction in one German blog: It enables "total surveillance."[39]) None of this thinking reflects what would be best for society. If technology could prevent all of us from doing stupid and dangerous things on the road, shouldn't we embrace it? We are balancing nothing less than life and death against our feelings of control.

While writing this book, I watched a tragic event unfold at New Jersey's Rutgers University. Freshman Tyler Clementi's roommate allegedly shared images of him making out with another man in their room. Clementi committed suicide by jumping off a bridge. The roommate and his friend, who reportedly participated, left school. One life was lost. Others were shattered. Because of my public opinions on the topic of privacy, *CBS Evening News* summoned me to be challenged by then anchor Katie Couric.[40] She pressed me again and again to say that the internet makes this story different, that the internet is the danger and it's the internet we should be teaching our children about. She sought a "teachable moment" about the net. I argued that this was a teachable moment about life. The internet surely adds speed, reach, and permanence. It can amplify mistakes. But the real lesson here is the same it has always been: the Golden Rule. The sin in that dorm could have been committed with a Kodak camera, a telephone, a letter, or a whisper. Society bears responsibility as well. That anyone would still feel shame about being revealed as gay and would make such a tragic decision is also our failing. If we think that technology is the problem, we risk ignoring the deeper faults and more important lessons.

A few months later, I spoke with a few dozen New York City high school students in an event at CUNY's Baruch College. They confirm boyd's and Pew's research for me. These teens manage their privacy settings. They know that anything they say just to friends on Facebook can be spread further by one gossipy classmate; they have learned by being burned. They're aware that college admissions officers and employers look at what they do online; to my surprise, they understand and don't mind it. A few of these students use tools such as YouTube to find a public for their creativity, but this group—including students from a school for the performing arts—are venturing into the public carefully. After the Tyler Clementi case, I ask whether the media are right, that the internet is causing an epidemic in bullying. Oh, there's bullying, they say, but their generation didn't invent it. The web only makes it more visible.

These matters of privacy always have and always will come down to a question of how people treat one another. We will try to encode that into our laws, rules, etiquette, and technology. But in the end, I wonder whether it is impossible to find a single meaning for the word "privacy." I've come instead to believe that privacy is an ethic. That's where I find my definition.

The Ethics of Privacy and Publicness

We may be looking at privacy through the wrong end of the telescope. Rather than viewing it from our perspective—as originators of private information and possibly victims of its violation—we can also see it from the perspective of the person, company, or institution that gains access to our information. The question is, once they have that information, what do they do with it? That is where the ethical choices are made and where the responsibility lies. When you've revealed something to even one person, the information is public to that extent. Whether it becomes more public is now up to the person you've told. When Steve tells Bob that he is getting divorced, Bob is the one facing decisions about what to do with that knowledge. Bob should ascertain whether he has Steve's permission to tell others. Bob should ask himself why he'd pass it on—to gossip and hurt Steve or to gather support around Steve and help him.

Publicness, on the other hand, is an ethic governing the source of information. If Sally has breast cancer, she needs to decide whether good could come from sharing that information. Would she inspire her friend Jane to get examined? If there seems to be a rash of cases of breast cancer where Sally works or lives, would her new data help pinpoint the source of the problem? Sally doesn't have to share. But by not sharing, she could be affecting others. The responsibility is hers.

Thus, privacy is an ethic governing the choices made by the recipient of someone else's information. Publicness is an ethic governing the choices made by the creator of one's own information. Or, put even more simply:

Privacy is an ethic of knowing. Publicness is an ethic of sharing.

I'll break down each of these tenets to see how well they apply to the choices we face and address the privacy concerns I've outlined above. First, the ethic of privacy:

- **Don't steal information.** Don't take information from people without their knowing it. Don't lie to get it. Don't snoop. Don't trick someone into giving it to you. Be open about having it. I'm not say-ing explicit, signed consent is always necessary—we don't want to fill in legal forms and disclaimers every time we say anything—but disclosure is a minimal requirement. People should be aware of the eyes on them.

- **Be transparent about what you will do with information.** When bloggers ask, "Is it OK if I blog that?" they start from a presumption of privacy in the conversation. Companies should operate similarly. Rather than being forced by regulators to reveal when they sell or pass data to third parties, they should make that intent clear up front. Without transparency, there cannot be informed consent and trust.

- **Protect information.** When you are entrusted with someone else's information, it is your responsibility to secure it appropriately. When, for example, through carelessness, you make shoppers' credit-card numbers vulnerable to theft and exposure or an online game network exposes users' email addresses and passwords, you lose the faith they place in you.

- **Give credit.** Taking and spreading information without giving due recognition to the source is another form of theft. Provenance mat-ters. The easiest way to give credit online is with a link.

- **Give people access to their own information.** People should be able to see what information is held about them. When possible, they should be able to correct or challenge it. Yahoo offers users a tool to adjust the ad categories aimed at them. Credit services are required to let us see what they say about us. Our information—including our creations (whether videos we make or lists of friends we compile or information about our purchases)—should be por-table. We should be able to copy and export this material (it is, after all, about us).

- **Don't use information against people (unless they deserve it).**
  Don't lie about people. Don't blackmail them. I hope that's obvious.
  I add the parenthetical exception "(unless they deserve it)" because
  criminal law and investigative journalism are based on using a person's own information against him to prove a misdeed or expose hypocrisy. That's particularly important if that person is in a position
  of power—for example, a sanctimonious legislator who writes laws
  against drugs but turns out to use them herself.

- **Context matters.** Though context can be hard to judge and inten-
  tion difficult to intuit, one should try. I told you earlier that I once
  wore an adult diaper. If you're going to repeat that, I hope you'd
  include the context of my prostate surgery so as not to give your
  audience the impression that I have odd urges.

- **Motive matters.** When you do reveal information about someone,
  ask yourself why you are doing so and why the person may or may
  not want it to be revealed. Are you doing so in the belief that it will
  help them or hurt them? Are you doing so to help yourself rather
  than them?

- **Add value.** When you use my information, it's best for both of us
  if you return value to me for doing so. In aggregate, Google adds
  value to our search data when it takes our links and clicks, analyzes
  that information, and feeds it back to us in the form of relevant
  recommendations. The wise company would find ways to use my
  data to tell me more about myself than I could know on my own: a
  credit-card company, for example, could tell me how my communications spending compares with other families'.

Note that these notions focus on the use of information, not on
technologies. We need principles that can adjust to any new tool. They
should apply to individuals, companies, and governments alike—though
the more one knows about another and the more one can do with it, the

heavier the responsibility (thus lovers, doctors, Facebook, Google, and government bear greater weight). Whether and how such guidelines are codified in laws and regulations—and how they are enforced—are questions that can be answered only after we first try to grapple with adapting society's norms in the new age. If we can't define privacy, how can we expect to legislate it?

What does the equivalent ethic of publicness entail? It includes aspects covered earlier in my exploration of the benefits of publicness—be transparent, be open, be collaborative, give respect, give value—and these:

- **Be generous.** If you have information that could in any way be valuable to others, you must ask yourself: Why not share it?

- **Share for a reason.** If you can't imagine why anyone would care about what you're sharing—whether a tweet about your breakfast or a press release about your product—don't share it. There's enough noise already.

- **Use common standards.** When sharing data, do so in a format that can be read and analyzed by many programs. Enable others to share the information and mix it with more information to learn yet more.

- **Protect what's public.** Recognize public information as a public good. Resist efforts to shrink public knowledge, and support expanding it.

Publicness is less complicated than privacy. It's not about fear, limitations, and laws. It's about sharing, connecting, joining, learning, acting, adding. Knowing one's privacy is secure makes it easier to be public. I hope we all will feel free to be more public.

How Public Are We?

We Have Met the Public, and They Are Us

"The data suggests that people are self-violating their privacy at a humongous rate," Google's Eric Schmidt says. "The number one cause of future privacy issues is going to be self-publishing of information. It's the sum of photos, blogs, Facebook, Myspace. . . ."

As I write, upward of 750 million people—approaching a tenth of the world's population—are sharing their identities, relationships, thoughts, photos, actions, likes, and lives on Facebook alone. They have an average of 130 friends and together share 30 billion pieces of content—photos, links, notes—each month. That's a billion acts of sharing a day.[1] As of early 2011, more than half of Americans over age twelve had Facebook profiles (up from 8 percent in only three years).[2] Nearly three-quarters of U.S. teens and young adults and half of U.S. adults overall used social networks in 2010.[3] More than 175 million Twitter users tweet about 100 million times a day,[4] adding up to 25 billion messages in 2010.[5] And a consistent tenth of American adults go to the trouble of maintaining a blog (14 percent—and dropping—of teenagers blog). [6]

Together, Twitter, Facebook, and bloggers are challenging Google as the primary path to content discovery online. Google sends 4 billion clicks a month to news publishers, while—in an admittedly apples-to-kumquats comparison—Bit.ly, just one of the services used on Twitter to shorten web addresses, causes more than 8 billion clicks a month (not all to publishers; some go to cat videos).[7] And Twitter is only a fraction the size of Facebook, another platform for peer-to-peer links. By sharing publicly, we people challenge Google's machines and reclaim our authority on the internet from algorithms. John Henry would be proud.

We share much more than links. We share our stuff. Flickr holds more than 5 billion photos, most of them public, and Facebook has up to ten times as many.[8] YouTube receives thirty-five hours of video every minute. Its videos are watched 2 billion times a day.[9] We also share our updates. Foursquare, Gowalla, and Facebook let us tell friends where we are. Goodreads lets us share the books we are reading, Last.fm the music we like, Delicious the web sites we bookmark, and Scribd the documents we create. There's even a service, SlideShare, that helps anyone publish PowerPoint presentations (now, *that* may be publicness taken too far). Perhaps the most extreme of all: Covestor lets investors share their stock trades (and profit from it),[10] and Blippy lets shoppers reveal their purchases to friends (more on that later). By the time this book makes it into print, every one of those numbers will have multiplied and so will the corps of services that help us share. Add it all up, Schmidt says—in one of those statistics that no one, not even Google, can confirm—and we create as much information *every two days* as we did from the dawn of man through the year 2003.[11]

The Pew Internet & American Life Project keeps invaluable numerical tabs on the growth of technology in society. Some findings from its 2009 and 2010 surveys:[12]

- Almost half of adult internet users search for people from their past, looking up old friends and lovers. When I give talks, I ask how many have searched for old girlfriends or boyfriends. Some hands go up, some don't. I congratulate the honest ones.

- 57 percent Google themselves. That is, most people expect to have a public identity, a Google shadow. Like Peter Pan, the majority discover their shadows: almost two-thirds find relevant material about themselves on the web.

- Almost half of online adults agree that getting to know new people is easier and better because we can look one another up online. 16 percent of internet users have Googled people they were dating (that seems low, doesn't it?), but 34 percent of online dating users go online to check out dates.

- 42 percent of internet users and 61 percent of bloggers are more likely than others to visit a park, belying the reports that bloggers are venomous vampires who spend their lives in PJs. A later Pew survey found that people online are more likely to be involved in volunteer organizations.

- 44 percent of online adults consult the web seeking information about people who perform professional services for them. Memo to doctors, lawyers, and manicurists: You'd better be online and public.

- Just 33 percent of internet users worry over how much information is available about them online. That seems low, given all the hoo-ha over privacy. And it's getting lower, declining by 7 percent in four years. So perhaps we're not in such a panic about privacy; media and government may be panicking on our behalf.

- More and more, we will go online not from computers but from phones. 81 percent of young adults get to the net via mobile phones. Note well that the phone will be a key device to bridge the digital divide between the connected and unconnected around the world as well as in the U.S. Pew says that African-American adults are "among the most active users of the mobile web" and that's growing faster than other segments.

- 86 percent of teen social network users post comments on friends' pages, and 83 percent comment on friends' pictures. These interactions occur in the view of others, making relationships a bit of a spectator sport.

- About a third of Americans create and share content—photos, video, art, stories. That is, a third of Americans make media.

How Public Is Too Public?

Oprah Winfrey made her career on the parabola of publicness. In the 1980s, soon after she became host of a morning show in Chicago, she copied—corrupted, some say—Phil Donahue's talk-show format by featuring real people humiliating themselves. A sampling of topics at the time: "Housewife Prostitutes," "Man-Stealing Relatives," "Unforgivable Acts Between Couples," and "Hairdresser Horror Stories."[13] Donahue, who was still dutifully and seriously exploring atheism, homosexuality, and public policy, got trounced in the ratings until he, too, dabbled in tabloid TV. The virus spread, spawning Jerry Springer, Ricki Lake, Jenny Jones, and Geraldo Rivera. They all exploited a primal urge in the audience: to be famous. Back when fame was hard to get, if you couldn't buy it with talent or money, many paid with humiliation. As a TV critic at the time, I remember commentators fearing the airwaves would be taken over by trailer trash. But then, in the 1990s, Winfrey experienced a Damascene conversion, deciding to use her power for good—and even better ratings. That is the arc of publicness: obscurity to overdose to balance.

   "One of the biggest things I've learned over the years is that people want to be heard," Winfrey said in a YouTube video as she announced the end of her broadcast show. "Every human, no matter what age, no matter how old we get, is looking for the same thing. What everybody wants is to know, 'Did you see me? Did you hear me? And did what I say mean anything to you?' "[14] On Twitter, Scott Heiferman, founder of Meetup, responded to Winfrey, saying we no longer need her and media to be heard. We have Facebook and YouTube instead. In a mastery of Twitter abbreviation, he wrote that the secret of the social net is "See+hear F2F IRL." Translated: See and hear, face-to-face, in real life.

   In 1996, Jennifer Ringley, a nineteen-year-old student, set up a camera in her Dickinson College dorm room and sent pictures to the still-newfangled web once every three minutes, showing anything that happened, including her drying off after showers and even having sex under the covers, though one had to wait a long time for those bits. It was hardly porn, only life made public. In 1998, she added video. In 2003, she turned off the camera, and today she lives in relative obscurity. Four years after Ringley quit, the art of what became known as "lifecasting" advanced. In March 2007, Justin Kan wired a camera to his hat and hauled connected computers on his back to broadcast his day, all day. He was televising not himself but instead everything he saw.[15] Later that year, Kan tired of the exposure and turned off his camera. He opened the platform he'd built, Justin.tv, as a service for others to use to broadcast live. One of its lifecasters, Justine Ezarik, calling herself iJustine, mixed the metaphor, broadcasting what she saw but also taking off her camera cap and putting it on the dashboard as she drove to the corner café or the Apple store. People wanted to see Justine, not just what she saw. All those artsy projects were captivating . . . for a while. Real life, it turns out, is predominantly dull. But Apple and its competitors watched those pioneers and soon offered the technology they had used, neatly packaged into phones with cameras on both the front and the back. Now we all can show the world what we see, and be seen.

We all live on the parabola: We discover something new. We get carried away with it, frightening pundits, who declare that life will never be the same. Then we find our balance. That's what we're doing now with many technologies as we find limits and test them, resetting the lines between openness and exhibitionism, utility and exploitation, credit and fame, self-respect and narcissism, protection and isolation.

Take cookies. Those lines of code record whether your computer has visited a web site. The next time you visit that site, it may use your cookie so it can target content and advertising to you, and so it won't show you the same ad over and over. It uses cookies to measure audience and traffic—how many users visited, how often. Cookies may be placed by the site you're visiting or by the company that serves its ads. Most cookies do not hold personally identifiable information (PII)—specifically, your name. They are usually anonymous. You can see which companies are using cookies and other technologies on the sites you visit by downloading a program such as Ghostery. You can also block all cookies. On most browsers, go to the preferences and find the privacy settings, where you can forbid cookies or erase them when you close the browser. Some browsers also let you open private sessions (called an "incognito window" on Google's Chrome) to prevent the browser from keeping any trace of your actions—except for your downloads. (Google still warns—with no discernable irony—to watch out for people standing behind you and "surveillance by secret agents.") You can install software to block ads. And, of course, when asked for information about yourself, you can lie (you're not the first wag to say you're ninety-nine years old).

There are consequences to donning the cloak of cookieless invisibility. Sites with passwords will require you to remember them and sign in again and again. You may get served less relevant content. You will not get targeted ads and will be more likely to see lower-value generic ads, which tend to be the most irritating kind, telling you to slap a dancing monkey or some such. You will become less valuable to the sites you visit. If your browser cookie shows you to be a user interested in, say, travel, your view of a travel ad could be worth 2.5 cents to the site showing it—more if you click on it. If you turn cookies off and get a generic ad instead, the revenue for that ad space could plummet to 0.015 cent or zero. You will affect the site's business and its ability to subsidize the often-free content or service it gives you. Induced by media hysterics over ad tracking—"Marketers are spying on internet users," The Wall Street Journal cries[16]—various government agencies have threatened to enact do-not-track lists akin to the do-not-phone list available in the United States. There's no real need, since users already have tools to stop tracking. This legislation would be like the government intervening to say you have a right to get a newspaper without ads when ads are what pay for the paper's content. But enacting do-not-track lists would be politically popular, since advertiser tracking is unpopular.

If I sound like an apologist for advertising, it's because advertising is the most visible means of support for journalism and media. I'll concede—and I'm confident we'll all agree—that most advertising sucks. It knows nothing about me and my needs and desires. It only shoves messages at my eyeballs. It is an intrusion and a waste of attention. Even so, without the ability to target, advertising would become even less relevant and more annoying as quality advertisers flee the web. The web, like the highway, would be filled with generic billboards. Worse, media companies could be forced to take their content and services behind pay walls—or die.

The advertising and media industries have themselves to blame for the fuss over tracking. They have not been at all transparent about what they monitor and how. They do not give users enough visibility into the data collected about them or enough control over the use, accuracy, and relevance of that information. They waited for regulators to come sniffing before they started trying to regulate themselves. It's probably too late for them to regain trust. Advertisers should be operating like Amazon.com, which makes it clear that it uses my purchases to target recommendations and also makes it easy for me to correct its misimpressions. That makes Amazon.com's marketing to me more targeted, relevant, and efficient and in turn makes me as a customer more valuable to the company. If ad tracking afforded similar benefit, transparency, and control for consumers, I think it could be tolerated.

The word "cookie" at least sounds benign. "RFID chip" sounds more menacing. In 2010, Walmart said it would attach these trackable chips to pants so it could monitor and restock inventory in stores. One self-proclaimed privacy advocate, Katherine Albrecht, raised alarms, which newspapers reflexively reported. Had they looked at the advocate's history—had they done their reporting—they would have found that she opposes the chips—as well as supermarket frequent-shopper cards—because she believes, quite seriously, that they are the "mark of the beast" and a sign of the End Times.[17] On this absurd basis, she raised media hackles, adding to a growing perception that privacy is under attack. But let's examine the danger. "Some privacy advocates," The Wall Street Journal says, "hypothesize that unscrupulous marketers or criminals will be able to drive by consumers' homes and scan their garbage to discover what they recently bought."[18] Are they serious? Just what would the villain do with that information? Look for homes to rob owned by people of the same size as the burglar? I suppose it's theoretically possible that through combining data, one might intuit that I've gained weight and my insurance company could raise my premiums. I'm willing to take that risk to ensure that the store has the right sizes on the shelf. I have no problem exposing my pant size: 33/34.

Esther Dyson exposes much more. She and nine other people have published their entire genomes online at the Personal Genomes Project.[19] As an investor in the DNA-mapping company 23andMe, she tells me she wants to make a point: that there's no harm in such transparency. She half thought that a pharmaceutical company would come to her with targeted drug suggestions. None did. Some friends wondered whether her family would object because her DNA has much in common with theirs. Dyson says they're all scientists and so she shrugs. When I tell her that by revealing my prostate cancer I am sharing not only my DNA but that of my male heirs, she grins and says, "Don't flatter yourself. Everyone gets prostate cancer."

I have used 23andMe. After spitting into a bottle and waiting a few weeks, I received a report on my DNA. It told me where my ancestors were from (northern Europe, hardly a surprise given my pallor). It also told me whether I stand a higher or lower than average genetic chance of getting diseases including type 1 diabetes (I have nine times the average risk, which makes sense since my mother has the disease); melanoma, rheumatoid arthritis, and stomach cancer (all higher odds); psoriasis, multiple sclerosis, and Crohn's disease (all lower odds). It doesn't say much about prostate and thyroid cancer and atrial fibrillation, all of which I have. 23andMe gives me the option to be contacted by people with whom I share genes on the chance of finding relatives. I would like that because on one side, my family tree disappears quickly up a holler in West Virginia. But so far I've found no kissin' cousins.

Once my DNA has been recorded, it *could* be used against me. If an insurance company can demand my health records, can it also demand my DNA map? Could it refuse me coverage based solely on my genes? Might an employer not hire me because I have a chance of getting a condition that could affect my work or raise its insurance premiums? Would police know to subpoena my DNA from 23andMe in a criminal investigation? I talked with Michael Fertik, founder of the privacy services company Reputation.com, who worried that potential mates could research not only each other's likes and dislikes on Facebook but also decide on their marriage worthiness on the basis of DNA. All these things and worse *could* happen. So what do we do? Forbid Dyson and me from revealing or even analyzing our DNA for our own privacy protection? Or, following danah boyd's rule, do we instead regulate the use of the data, forbidding insurance companies and employers to discriminate on the basis of DNA? The problem with regulating this new technology around the bad things that could happen is that it also cuts off the possible good. 23andMe quizzes volunteers about their health histories to correlate genes and ailments, which could be useful in finding causes and even cures for a condition. This knowledge could save lives if we are willing to share it. Where should the line be? We don't know yet.

Oversharing

After revealing much about my penis in public, I heard only three complaints. One came from author Mark Dery, who had often been critical of me and my opinions anyway. He took me on for oversharing.

Oversharing. Odd word. How much sharing is too much? How much is enough? Who's to say? Emily Gould is credited with inventing or popularizing the term when she wrote in The New York Times Magazine about her oversharing while she was a writer for the gossip blog Gawker. There, Gould covered celebrities and became a bit of one herself.[20] She didn't so much define the word "oversharing" as live it, telling readers about her boyfriends and breakups and turning the web, as she put it, into group therapy—until she thought she'd gone too far. "I had made my existence so public in such a strange way," she writes, "and I wanted to take it all back, but in order to do that, I'd have to destroy the entire internet. If only I could!"

Now Dery was accusing me of oversharing. He was reacting to a Time magazine essay, "In Praise of Oversharing," by Steven Johnson, who tells the story I've told you. "Jarvis is a friend of mine," Johnson says, "but it may tell you something about the strange mediated state of 21st-century friendships that I first found out about his cancer diagnosis in a Twitter update he sent out linking to his original blog post. This is how we live now: we get news that we're facing a life-threatening disease, and the instinctive response is, *I'd better tweet this up right away.*[21] As Johnson puts it, "We are overexposed to overexposure." But at the end, he comes around to my way of thinking: "We habitually think of oversharers as egotists and self-aggrandizers. But what Jarvis rightly points out is that there is something profoundly selfish in not sharing."

Dery is not convinced. On his blog, he argues that oversharers like me have a "disease of the psyche" marked by "an obsessive need to connect," that we are "redrawing the boundaries of publicly acceptable behavior," and that I emblemize "the blogorrheic, tweet-expulsive times we live in, when so many of us feel the need to broadcast our every thought, at every minute, to everyone."[22] He worries that we are "dissolving the membrane between private 'I' and public self" and "reversing the polarities of public and private."

What is it to Dery if I talk about my cancer? I think he reveals more about himself than I do about myself, even if I'm the one with the exposed life. "Is the desire to broadcast the most mortifying details not only of our private lives but of our private parts really about the desire to Feel the Love on an epic scale?" Dery asks. "If so, isn't it selfish rather than selfless?" In his view, cancer and penises are mortifying. "Exhibitionism," he adds, "is a form of social dominance." How? Dery is the one trying to tell me what not to say. Only later in our public jousting does he mention that he is a prostate cancer patient himself.

Note Dery's verb: "broadcast." He puts this discussion into mass-media terms. "This," he says, "is partly about the media-age article of faith that nothing is really real unless it's recorded and, increasingly, shared." He says my talking about my cancer has "everything to do with our media-age fixation on fame." Oh, yes, I respond, I want to become famous for my limp and leaky dick. Dery accuses me of trying to emulate mass media when I'm merely living out loud. He's the one who sees life through a mass-media lens, thinking of the internet as a TV station rather than as a means for real people to connect with one another.

Because he thinks we are on camera, conversation to him sounds like performance.

There is, no surprise, a web site devoted to oversharing, filled with tweets. "Have such volcanically deep zit roots in my chin that it feels like someone hit me with a right cross," says one at Oversharers.com. More: "Love means rubbing Desitin on your husband's ball rash." "I just burped loud enough that the dogs downstairs started barking." "Hello PMS and your friend Cravings which has just made me consume ¼ lb of salami in the car on the way home from the grocery store." "The gentleman using the urinal next to me showed me the kidney stone he passed, then took it back outside to the party." "Man. That Chipotle ran thru me like a track meet."

Julia Allison is accused by her critics of oversharing. She is a New York gadabout who invites you to follow her on Facebook, Twitter, Tumblr, YouTube, and Myspace and at JuliaAllison.com. She wrote a dating column when she was in school at Georgetown and then came to New York and ended up working or appearing in much media there. In The Times, Emily Gould measured her own oversharing against Julia Allison's: "I was initially put off by Julia's naked attention-whoring—'Attention is my drug,' she often confessed. In thousands of photos on her Flickr feed she posed, caked in makeup, like a celebrity on the red carpet, always thrusting out her breasts and favoring her good side. But in the midst of this artifice she was disarmingly straightforward about how badly she craved the attention that internet exposure gave her—even though it came at the expense of constant, intensely vitriolic mockery."[23]

Allison attracts detractors who go so far as to maintain web sites to shoot snark at her.[24] "I am a media personality, for lack of a better term," she tells me onstage at a New York Internet Week event, "and a professional sharer—not oversharer." She stiffens at the term "oversharer." "It is a distinctly pejorative term," she says, for it implies a lust for fame. When you tell your friends about your life, they don't accuse you of seeking fame, she argues. "People bully me on the internet," she complains. Does she feel burned? "Yes, yes, I do." Does she feel trapped? "Yes, I absolutely do feel trapped by this. I envy people who can just change their privacy settings on Facebook." Why not quit and become a nun? "I continually play around with that. I went to an ashram a month ago and went completely off the web," she reveals with no glint of irony. Yet she is self-aware when I ask her why she gets attacked. She smiles and concedes, "Oh, I got a little bit of an ego."

Allison wishes the nastiness online could be controlled, but she's not sure how. "I'm all for freedom of speech," she says, "but I think we need to talk about something else. We need to protect people. . . . It's a gray area between defamation and harassment." She wants rules. She wishes someone had told her when she went online, "This is what you will be giving up. There were certain consequences that occurred that I never would have foreseen." But she acknowledges that some good comes from living in the open. "The one thing the internet has assured: I cannot lie about anything," she says. "I have hair extensions. I plan on getting Botox."

Allison and I happen to share one (well, at least one) detractor online. I tell her I gave up looking at what he has to say. I blocked him on Twitter and don't read his site. So the only time he crosses my line of sight is on the rare Twitter search. And I ignore him. I know what he'll say and don't care. I won't give him the pleasure of reacting. Law of the playground and number one rule of internet interactivity: "Don't feed the trolls." Allison says others' barbs have cost her jobs, relationships, and friends. Yet she keeps sharing. Why? "I can pay my rent because of sharing on the internet." It's a living. Sharing is her business.

The Public You

Identity and Reputation

I'm asked sometimes whether I believe in radical transparency. "I'm still wearing clothes, aren't I?" I respond. Then I add, "You're welcome." Technology need not strip us naked. Whether you share some revealing bit on Facebook or Twitter is up to you and is in your power. How will you know when you've gone too far? What *is* oversharing? I think it's sharing that which you regret having shared. Oversharing, then, is not in the eyes and ears of the beholder. It is in the mind and mouth of the sharer. If you don't want the world to know it, your safest bet is not to say or do it. As Eric Schmidt infamously said in a 2009 CNBC interview, "If you have something that you don't want anyone to know, maybe you shouldn't be doing it in the first place."[1] Call that Schmidt's commandment. Like other Schmidt quotes, it raises hackles. He's right, though.

If you trust a secret to a friend who shares it, your problem could be your choice of friends. The fault, then, is not with the technology but with us. Computers don't embarrass people; people do. Don't we all operate by the email rule that if you don't want what you're saying forwarded to the wrong person, you probably shouldn't say it in email? We've all had that lesson. We're burned. We learn. We adapt. Schmidt says there was a time when mothers didn't reflexively hold their children's hands when they crossed a path. Then came new technologies—horse-drawn carriages and cars—and behavior changed. Today's technologies are, of course, more complex. "I don't believe society understands what happens when everything is available, knowable, and recorded by everyone all the time," Schmidt tells The Wall Street Journal.[2] He tells me, "When I was growing up, was I lonely? Sure. Was I bored? Sure. Did I not know everything? Sure. So now you're never lonely because your friends are always reachable. You're never bored because there's infinite streams of information and entertainment. That is a fundamental change in life." Technology leads to new choices, which ultimately leads to more control. We get to say more. As a result, we also hear more. That means we need to examine our norms about how we react when other people are more exposed.

In 2010, someone leaked the fact that Lori Douglas, a senior judge in Manitoba, had once had pictures of herself in acts of bondage and oral sex on a web site devoted to interracial couples—before she rose to the bench. The problem for her came when critics questioned whether, by not revealing her sex life, she had responded accurately to this question when she was being considered for a federal judgeship: "Is there anything in your past or present which could reflect negatively on yourself or the judiciary and which should be disclosed?" Who among us could leave that blank? Who is fool enough to fill it in? We all have embarrassments. If we have no foibles, how qualified would we be to judge the foibles of others? A columnist for The Globe and Mail called for Douglas' resignation because "she should have known better."[3] But the Montreal Gazette asked, "Do we expect our judges to be plaster saints? . . . In 2010, are most Canadians really offended and morally repulsed by interracial bondage sex? Should they be? The world is changing. Social norms are changing."

Bondage on the bench is hardly a typical situation. But then, this debate tends to be held around the extremes. In *The Future of Reputation,* Daniel J. Solove gives a reasoned and reasonable approach to reputation but still uses edge cases to make his point: a girl whose picture was passed around Korea after she let her dog poop on a train and refused to clean it up; a flasher in a New York subway whose picture ended up in newspapers while he ended up in jail; a young man who was once in prison and whose history haunts him on dates; the "Numa Numa" and *Star Wars* kids who posted videos online and became viral stars; a blogger writing about her sex life on Capitol Hill who garnered media notoriety and a book deal. Edge cases are good at feeding debates but not at informing norms.

The question in this chapter is how the average person—you or I or the random person I see on the train who's keeping his dog in line and his pants zipped—can and should cope with publicness. How do we balance the risks and benefits of posting a picture to Facebook or Flickr, stating an opinion in a blog comment or forum, or telling friends what we've bought or where we are? How can we establish and manage our identities and reputations—and can we do so at all?

One tactic to cope with the fear of exposure and overexposure is anonymity. Anonymity has its place. It protects the speech of Chinese dissidents, Iranian protestors, and corporate truth tellers. It protects the gay teenager who needs to talk about his life but dares not reveal himself in school. It lets people play with new identities. When the game company Blizzard Entertainment tried to bring real identities into the forums around its massive, multiplayer games, including World of Warcraft, players revolted—and no wonder: Who wants everyone to know that in your other life, you see yourself as a level 80 back-stabbing night elf rogue who ganks lowbies at the Crossroads?[4] Taking on identities—pseudonymity—is the fun of it.

But anonymity is often the cloak of cowards. Anonymous trolls—of the human race, not the Warcraft breed—attack people online, lobbing snark at Julia Allison, spreading rumors and lies about public figures, sabotaging a politician's Wikipedia page, or saying stupid stuff in the comments on my blog. I tell commenters there that I will respect what they have to say more if they have the guts to stand behind their words with their names, as I do.

Real identity has improved the tone and tenor of interaction online. That was Facebook's key insight. Twitter's, too. Tweeters want credit for their cleverness; they are rewarded with followers and retweets, their nanoseconds of microfame. Facebook is built on real relationships with real people in real life. "That doesn't mean that every single thing is true," Mark Zuckerberg says. "But on balance, I think it's a lot more real than other things on the internet." Zuckerberg believes we have one authentic identity and says it is becoming "less and less true" that people will maintain separate identities. Emily Gould, admitted oversharer, agrees. Julia

Allison, on the other hand, sides with those who say we need to maintain many identities—one for work, another for school, another for home, another for friends. Those folks say we get into trouble online when these identities mix and blur, when our boss sees our picture from the college beer party (as if bosses never drank beer). In a New York Times Magazine piece arguing that "the Internet records everything and forgets nothing," Jeffrey Rosen tells the story of a twenty-five-year-old student-teacher who was deprived of her diploma after posting a Myspace photo of herself drinking, over the caption "Drunken Pirate."[5] On his blog, Scott Rosenberg counters that "the photo is harmless; the trouble lies with the people who have turned it into a problem."[6]

What needs to change is not so much our behavior, our rules, or our technology but our norms: how we operate as a society and interact with one another. When presented with someone's public face, which may differ from our own, is our response to disapprove, condemn, ridicule, and snipe, or is it to try to understand differences, offer empathy, overlook foolishness, and share in kind? When we disapprove—and we all sometimes do—we can be guilty of intolerance. When we offer empathy, we become open-minded. Because we are all more public, I wonder whether we will soon operate under a doctrine of mutually assured humiliation: I'll spare you making fun of your embarrassing pictures if you'll do the same for me. "An age of transparency," says David Weinberger, author of *Too Big to Know,* "must be an age of forgiveness."

There are two forces at work here: identity and reputation. Our identities are the first-person expressions of ourselves. Our reputations are others' third-person views and conceptions of us. Thanks to our increasing publicness, the two are coming closer and sometimes into conflict. As I was discussing these topics on my blog, Weinberger left a sage comment wondering about what he called the private-public axis:

Marilyn Monroe was a public figure, but most of us are private citizens. That used to be pretty easy to compute, and, because of the nature of the broadcast medium, it used to tend toward one extreme or another: He's Chevy Chase, and you're not. But there's another private-public axis: who we really are and how we look to others. We have tended to believe, at least in the West, that our true self is the inner self. The outer, public self may or may not reflect our inner, private self, and we have an entire moral/ normative vocabulary to talk about the relation of the two: sincerity, authenticity, integrity, honesty.[7]

Those are the two identities we are trying to manage—not our work selves and our home selves, not our party selves and our serious selves, but our inner, real selves and our outer, show selves. When our inner and outer selves get into conflict and confusion, we appear inauthentic and hypocritical. In all our spoken fears about privacy and publicness, I think this anxiety is the great unspoken fear: that we're not who people think we are and we'll be found out.

These are new skills for everyone, celebrity and commoner alike. Marilyn Monroe never had to deal with blogs and Twitter, let alone twenty-four-hour TV news. She had press agents to create and manage her identity and security people to keep scary strangers away. Today, stars and pols have to deal with constant exposure. When we catch them in a contradiction of words or deeds—and it's not hard to do—they suffer the gotcha. Then again, stars such as Ashton Kutcher, Lady Gaga, and Howard Stern are grabbing the opportunity on Twitter to interact directly with their publics without scripts or PR people and reporters in between.

When people do find unflattering results in Google searches for their names, Reputation.com suggests the solution is not to hide but to publish more about yourself so that will rise on Google. The way to affect your reputation is often to share more, not less. The best solution is to be yourself. If that makes you uneasy, talk with your shrink. Better yet, blog about it.

Public Advice

I ask my nineteen-year-old son, Jake, to tell me the advice he would give his fourteen-year-old sister, Julia, as she enters the social whirl of high school online. He shrugs, not just because he is a teenage son but also because it all seems so natural to him. His generation—and in internet years, generations come more quickly—create their own norms around social tools. They use them differently than their elders do. When I first got onto Facebook with a university .edu email address, I asked Jake about the etiquette of friending. He told me it is impolite not to friend someone you know at school. My business peers tend to friend just the people they know well. Some see friending as an endorsement rather than merely a connection. Young people share jokes using Facebook's language and structure. They get divorced with the speed of a click. We adults need lawyers.

So it would be difficult to present a single set of concrete and specific rules and tips that will help everyone cope with greater publicness and less privacy. It's not just that the tools change rapidly. The tools also change society rapidly. Perspectives, expectations, even metaphors and analogs expire. (What is a phone when you are young and never make calls with it? What does it mean to "publish"? What does "friend" mean now that the noun is also a verb?) At the same time, society's norms become a rapidly moving target. Compare the sins and virtues of TV's *Mad Men* with the culture of today's internet. Look at how far we've come from single beds in Lucy and Ricky's apartment to the magnificent diversity of relationships on *Modern Family*. Imagine having a discussion about being unfair to bondage-loving judges only a few years ago, even in tolerant Canada. And now imagine it all sped up by the public web.

When sharing, please don't judge your success or failure in mass-media terms. The goal is not to get a million followers. Publicness starts in pairs. Who in her sane mind wants a million people watching all the time? I got an email from a young teenager who had started blogging and was trying to decide how to get more page views—by writing about teddy bears or skateboards. I told her to ignore that and write about what interests her, to explore and learn. The reward of publicness is best found in terms of relationships: making friends, reaching people with shared interests and needs, solving your problem or someone else's, finding out you're not alone, accomplishing something, having a laugh. Those are all good reasons for being public, at any scale.

Even as much as I celebrate publicness and am an optimist about its impact on our lives, I agree with the cautious, who say that we need to stop and think—and teach our children to stop and think—before exposing ourselves. Media literacy—or, as the Germans more aptly call it, media competence—involves more than consuming content. It now means teaching people how to make content and what its consequences are. There is a growing difference between conducting relationships in the public of one's small village and doing so in the global village. Information, both good and bad, spreads wider, faster. Just as the benefits can be greater, so can the costs increase.[8] So it's worth taking time to consider a few suggestions for the public life:

The tattoo rule. Blippy cofounder Philip Kaplan repeats this sage advice: Anything you put online is a tattoo. It's permanent. It won't go away. The web remembers. People may give you slack, but you can't be assured they will. What's seen as a sin today—drinking, smoking, smoking pot, snorting cocaine, or having sex on the side, to list a litany of presidential peccadilloes that have sparked less and less outrage over the years—may become accepted or even passé tomorrow. Or they may not.

Eric Schmidt says that young people should be able to change their names and start over at age twenty-one. He's joking. I heard him deliver this quip at the 2007 Personal Democracy Forum, where the audience laughed.[9] But when he used the line again in 2010, the irony was lost. A tempest broke out in the media teapot, as The Wall Street Journal reported Schmidt's quote and either didn't get the joke, or chose not to. "He predicts, apparently seriously, that every young person one day will be entitled automatically to change his or her name on reaching adulthood in order to disown youthful hijinks stored on their friends social media sites," wrote Journal reporter Holman W. Jenkins, Jr.[10] The meme spread like an oil slick. "Death by Twitter," Schmidt calls it. When he talks in public, people tweet a quotable line but "omit the preamble and the postamble and the context." He was not serious about all of us changing our names. "It was a joke. It is a joke. It will remain a joke," he tells me. His own gag demonstrates the rule: Say it once, and you've said it forever. But that fear should not force us to abandon our voices.

The front-page rule. It has long been said that you shouldn't say anything online you wouldn't want to see on the front page of The New York Times. In the age of WikiLeaks, that's truer than ever. But I'll also turn this rule around: You *should* say some things so they *do* get on the front page—the new front page of Twitter, Facebook, Google, etc. You never know when something you think and then share could click with others and have an impact. The new American dream is to go viral. It happened to me when I complained about my Dell computer on my blog. I was not influential, but my "Dell hell" message resonated with others. Because of the storm, Dell changed its ways and its relationship with customers.[11] So think about the front page when you share, not just out of fear but also with an eye toward opportunity.

The social-bankruptcy rule. The problem with getting connected to more people is that more people can bother you. They have expectations. They want you to answer their comments and questions and even snarks in email and now also in Facebook and Twitter and new services such as Quora, where users ask each other questions. Enough! I have long since declared email bankruptcy. I cannot answer every email I get even from the people I know, let alone from my new social-media friends. I am nostalgic for the days of the busy signal, when my time and attention did not appear to be abundant. To put the issue into technical terms: publicness doesn't scale well.

Technology might help fix the problems it causes. Facebook's News Feed algorithm tries to predict which updates will interest us most based on "how many friends are commenting on a certain piece of content, who posted the content, and what type of content it is."[12] Google Gmail tries to declutter email by killing spam and its Priority Inbox and ranks our email for us. We can imagine how new technology will change our customs. Once upon a time, when we were late with an obligation, we'd claim it was "in the mail." Then we'd claim that someone's unanswered message got "caught by the spam filter." Now we'll say, "I guess Google didn't think you were a priority." Marissa Mayer, a Google vice president, talks dreamily of the day when an algorithm will sort through all our feeds—email, news, tweets, Facebook updates—and deliver a prioritized "hyperpersonal news stream." Technology can't do it all, nor would we want it to. But it helped create this mess, so let's hope it helps fix it.

The don't-feed-the-trolls rule. More contact with more people also increases the odds of crossing paths with a jerk. The internet doesn't make jerks. It just makes them easier to see and gives them a megaphone. When you see a troll, don't feed it. They come into online conversations solely to provoke. Respond, and you give them what they want: attention and an opportunity to keep the attack going. If they come in to despoil a space you control—such as comments on a blog—erase their droppings. I don't advise cutting off people who simply and civilly disagree with you; that will hurt your reputation worse than the trolls would. But other members of your community will thank you for not encouraging bozos and evicting them.

The Cabernet rule. Certain potions can turn even the nicest, sanest person into a temporary troll. It has happened to me. After a long day and a tall drink, one's tolerance for stupidity is diminished and the odds become greater that one may call a fool a fool and live to regret it. Friends don't let friends blog, tweet, update social statuses, pose for Facebook pictures, make YouTube videos, or operate the heavy machinery of the internet when drunk.

The honesty rule. When you make a mistake, own up to it. People will trust you to do so the next time it happens. Blogging taught me not to erase mistakes but to cross them out so people know that I messed up, fessed up, and have nothing to hide. I've had to correct factual errors. I've also confessed to being wrong about the consequences of supporting the Iraq war, which was harder. It's not easy, but I've learned that corrections don't diminish one's credibility, they enhance it.

The Golden Rule. What's true in life is true online. Share information if it could help others. Share credit. Link. Linking shows your work and enables your readers to see your source and judge it for themselves. Share attention, too, by passing on good things you find. The retweet in Twitter—repeating something you've read and giving credit—is a substantiation of the sharing society. Make generosity a reflex.

The don't-be-a-fool rule. This rule supersedes all the rules above.

When I'm asked for the ideal corporate guideline for blogging and social media, I say it comes down to one thing: Don't be an idiot. The Labour candidate for Parliament who called a veteran Labour MP "a fucking idiot" was being an idiot.[13] The advertising exec who tweeted that he couldn't stand Memphis the night before going there to present to Federal Express, a company proud to be headquartered in that fair city, was being a fool.[14] The Washington Post sportswriter who published a false news report in Twitter hoping to entrap bloggers into linking to it was being a doofus.[15] The woman who tweeted about her job offer at Cisco and her weighing the "fatty paycheck" against "hating the work"—only to be caught by an employee who told her, on Twitter, that "we here at Cisco are versed in the web"—was being a knucklehead.[16] The juror who declared on Facebook before her case began that it was "gonna be fun to tell the defendant they're GUILTY" was being both ungrammatical and a dimwit.[17]

The internet is life, only bigger and faster. The lessons you learned as a child and those you teach your children about how to treat others all still apply. The net is still just a place filled with people.

The Sharing Industry

The Public Economy

"Privacy was once free. Publicity was once ridiculously expensive," says entrepreneur Sam Lessin. "Now the opposite is true: You have to pay in a mix of cash, time, social capital, etc. if you want privacy."[1] You pay for privacy in the effort and hassle it takes to manage privacy settings. You also pay in the opportunity lost if you choose not to be public and social. On the other side of the ledger, you can be rewarded—with attention, influence, information, deals—if you reveal yourself. This new economy tilts toward publicness.

There is money to be made in privacy. At more than one conference, I've watched the blossoming of a regulatory/industrial privacy complex. At the 2011 Reboot Privacy and Security Conference in Victoria, British Columbia, a vendor selling web security software to companies went through his PowerPoint reciting some of the statistics I listed earlier, but with a darker tone. He announced that hundreds of millions of people were using Facebook. He paused . . . dramatically. Then he said, "Scary." Why is that scary? He didn't say. He didn't think he had to. Scare and sell is his strategy. A government regulator took the stage and demonized Facebook's Zuckerberg and Google's Schmidt, whom she misquoted, telling the audience that Schmidt had dismissed privacy as irrelevant. She bragged about making more regulations and adding more staff. Scare and spend. I met the head of an association of chief privacy officers. Bet your membership is growing, I said. By the thousands, he replied. Scare and grow. There are venture capital funds targeting opportunities in the privacy industry. Scare and invest.

I've seen the same dynamic play out at the MediaBistro.com Digital

Privacy Forum in New York and the 2010 Privacy Identity Innovation (PII) conference in Seattle, where I dragged son Jake. A conference participant tweeted that two days at the meeting "makes me realize that the Internet safety talk w/my kids will be just as important as the sex talk."[2] True enough. But if our kids know more than we think they do about sex, they certainly know more than we do about the social internet. My son couldn't take all the hand-wringing. We escaped to lunch and he told me these people see problems where they don't exist. Show me how the world is falling apart, he demanded. He wanted to ignore them. I cautioned him that they could implement regulations that would change how the services he loves operate or make them too expensive to continue. I heard regulators there muse about extending the protections of COPPA—America's Children's Online Privacy Protection Act—past thirteen years of age to eighteen. That would mean services like Facebook would need parents' *written* permission to hold any personal information (such as names and locations) even for high school students. What it would mean in practical terms is that companies would avoid building services for teenagers—that's why there are so few for preteens. Privacy regulation will have a big impact on business. Privacy is also becoming big business. At these conferences, I ran into the same players: advocacy organizations that raise concerns and raise money; legislators who make laws and bureaucrats who make regulations; consultants who help companies comply with said regulations; and services such as Reputation.com, which promise to protect privacy. There's money to be made in privacy.

But there is even more to be made in publicness, in building the platforms that fuel sharing. These companies will have an even bigger impact on business and society as they shift power and unseat old institutions that still depend on scarcity and control. They include Blogger, Word-Press, Google, Facebook, YouTube, Flickr, Twitter, Blippy, Foursquare, Quora . . . and the list keeps growing. Here are four entrepreneurs who are building that industry, their creations progressively more daring and perhaps, to some, more dismaying.

Evan Williams: Blogger and Twitter

It's the rare person who gets to change the world and witness the impact of his legacy. Evan Williams has changed the world twice so far. He co-created Blogger, the service that popularized the form, permitting anyone to easily and instantly publish to everyone. It disrupted the structure of media, journalism, and, one could argue, political, commercial, and social power. He next cocreated Twitter, the impact of which is just beginning to be seen, though still not understood. Twitter has been credited with at least supporting roles in revolutions and with taking another quantum leap in the restructuring of media and marketing. Williams is a founding father of the sharing industry.

Like many other inventions of the digital age, each of Williams' cocreations was a by-product of another idea, practically an accident but for his instinct to sense and exploit their popular appeal. Pyra Labs, founded by Williams and Meg Hourihan, set out to make productivity tools. A note taker they built in that process turned into Blogger. It was by no means the first blogging tool. Many people credit Justin Hall with being the first blogger in 1994. Dave Winer created one of the earliest blogs, Scripting News, and also some of the earliest blogging tools. Jorn Barger coined the word "weblog" in 1997; Peter Merholz shortened it to "blog" in 1999; Williams reportedly came up with "blogger" after his service started in August 1999.³ So he didn't invent blogging any more than Bill Gates invented PCs. But he made it simple, cool, free, and popular. He stuck with his vision even after his money ran out, more than once, and his staff and partners left. Continuing to run the company out of his kitchen, Williams understood its potential.

I didn't. When Nick Denton—later the founder of the successful blogging company Gawker Media—first showed me Blogger, I shrugged. "So what?" I asked. "So you put some text on a web page. What are you so excited about?" I'd read lots of blogs. And when Pyra Labs was on its last legs, Denton got me to get my then-boss, Steve Newhouse, to invest a relative pittance in the company and keep it going. Newhouse was the rare old-media executive who grasped the potential of interactivity. But I didn't begin to comprehend the importance of blogs until 9/11, when I started writing my own, Buzzmachine.com, after witnessing and surviving the attack on the World Trade Center. Blogging soon taught me the power of the link and the benefits of publicness. I came to respect and appreciate Williams' determination to keep Blogger alive even as money ran dry and servers choked. The story has a happy ending: In 2003, Williams sold the company to Google, which still operates it.

Ev's grand irony is that he's quiet and unassuming—like a shy, wry Mr. Rogers. He does not seek attention as the millions of users of his tools do. On his blog, Evhead.com, he wrote nothing for more than a year—from December 2009 until March 2011, when he announced that he was leaving his full-time role at Twitter to go to a blank whiteboard and work on what's next. Though he has 1.3 million people following him on Twitter—at @ev—he has tweeted only six thousand times. I started a couple years after him, and my oeuvre adds up to three times more. The most prolific tweeter I know, NPR's Andy Carvin (@acarvin), has almost ten times Williams' output. "Your Twitter stream is who you are," a Twitter colleague says when I sit down to talk with Williams. "It's just who he is."

Before Blogger and Twitter, we had tools to make and duplicate content: carbon paper, invented in 1806;[4] Chester Carlson's Xerox photocopier in 1938;[5] the dreaded Kodak camera of 1890; Kodak's 8 mm home movie camera in 1932;[6] and, using patents registered by Thomas Edison in 1876 and 1880, the mimeograph that powered the publishing of political tracts and so-called zines for more than a century.[7] But our mass passion to publish and broadcast did not explode until Blogger. Why? It could be that it's so easy to use. It could be that the net added a simple means of distribution to our simple means of content creation. Yet with the web's debut in 1994 came Tripod, one of the first easy tools to create what we then called home pages.[8] Tripod didn't produce a wealth of publishing and social change. Mostly, it produced homely, forgettable, and soon-abandoned web pages. What was it that made Blogger change the world five years later?

Blogging opened the door to a new kind of writing, just as the printing press led to the invention of another new and similar form, the essay. In the French Renaissance, Michel Eyquem de Montaigne broke free of scribal text's reverence for ancient wisdom and made writing contemporary and personal. Starting in 1572, alone in his tower library, Montaigne began to answer questions he posed to himself. "In place of abstract answers, Montaigne tells us what *he* did in each case, and what it felt like when he was doing it," writes Sarah Bakewell in her book on Montaigne, *How to Live*. "He tells us, for no particular reason, that the only fruit he likes is melon, that he prefers to have sex lying down rather than standing up, that he cannot sing." What does that sound like? Bloggers! Twitterers! Those damned fools make their private lives public, and the public inexplicably reads it. The printing press brought out that instinct in Montaigne. A half-millennium later, Williams' tools brought it out in millions more.

There's another factor that ties these tools of publicness. Montaigne's *Essays,* Bakewell says, is "much more than a book. It is a centuries-long conversation between Montaigne and all those who have got to know him: a conversation which changes through history, while starting out afresh almost every time with that cry of 'How did he know all that about me?'"[9] A conversation. That's what blogging is, as I discovered the moment I blogged about my memories of 9/11 and then others linked to what I'd said and I linked back to them. We held a conversation in different times and places, among friends and strangers, all powered by links, in public.

"Blogger gave them a canvas that they didn't have before," Williams says. "Publishing on the web prior to that was a technical task, and Blogger turned it into a creative pursuit. . . . You could set up your own public and people can gather around if they care about it. And that was freedom. That empowered people to do whatever the heck they wanted." Blogs, he says, are like going to someone's house, where your friend is the host, and then she can come to your house, where you are the host. Blogger, that is, gave us our own spaces rather than forcing us to gather in a common space (such as AOL chat rooms or newspaper forums). The public became the accumulation of us and our spaces rather than a square someone else—government, media, schools—provided for us. Williams also takes pride in blogging's role in freeing the public from being "just consumers of news to actually thinking about it and analyzing it." Some laughed when we sat down to our keyboards. Technology's early adopters, Williams recalls, sneered at blogging as an instrument of narcissism. Journalists sniffed at the ramblings of amateurs. Companies regarded them as buzzing mosquitoes. But we know now that the introduction and popularization of blogging had profound impact.

Nobody thought Twitter's impact would be similarly profound in 2006, when Jack Dorsey came up with the idea while working with another partner, Biz Stone, at their podcasting company, Odeo. Dorsey wanted a tool to let small groups of people communicate via SMS (which is what sets its limit of 140 characters). At the South by Southwest conference in Austin in early 2007, Twitter took off. Attendees used it to share bon mots they heard, recommend sessions while they were under way, and meet up with friends. Two years later at SXSW, I watched a Twitter revolt bubble up as a ballroom full of geeks got upset when an interviewer hogged mic time from her subject, Mark Zuckerberg. Unbeknownst to the duo on stage, anger was building up in the so-called Twitter back channel. It erupted out loud when Zuckerberg cracked that his questioner should ask a question and the crowd cheered. I've seen a similar phenomenon in the classroom, where Twitter gives students their back channel, ending the teacher's monopoly on attention.

After Twitter's debut, Williams, Dorsey, and Stone dumped podcasting and shifted their attention—pivoted, as tech companies call a strategic turn. The new tool still seemed like a toy. I remember showing Twitter to my faculty colleagues for the first time. "OK, Jeff," one of them challenged me, "I understand why this is cool. But what could it possibly have to do with journalism?" We began to list how journalists could use it to find stories people were talking about, ask for information, and promote their work. She was soon sold. Alan Rusbridger, editor-in-chief of The Guardian in London, delivered more than a dozen reasons why Twitter matters to journalists. It's a useful list that applies to many organizations: [10]

- It's a form of **distribution**. Twitter links send billions of clicks to sites each month. The Guardian uses it to promote its stories and bring in new readers.

- It's an **early-warning** system. When an earthquake hits or a plane lands on a river aside Manhattan or a revolution boils up in the Middle East, people share what they're experiencing on Twitter. They're not trying to publish news. They're telling friends what's happening in their lives. NPR supertweeter Andy Carvin says he finds witnesses to such events by searching for phrases such as "holy shit!," "OMG!," and "WTF!" Smart companies also use Twitter to hear about problems with their products before normal channels can.

- It's a **reporting** tool. Williams sees Twitter users as reporters, "information-collecting nodes, millions of them around the world that are reporting back about what's happening around them." He envisions a day when Twitter users will be able to "see what's happening anyplace." I see enterprising journalists use Twitter to ask readers what they know and what they want to know. One of the best at utilizing Twitter is New York Times columnist Nicholas Kristof—@nickkristof. During Egypt's uprising at Tahrir Square, Kristof took to Twitter to warn, "Mubaraks thugs out here holding clubs with nails, better lookout." "It holds dictators accountable," he said on CNN.[11] "And one reason we're not seeing more atrocities in the Middle East, frankly, is that everybody knows there are people tweeting, there are people taking photos, and the word is getting out."

- It's a form of **marketing**. Dell uses Twitter to distribute coupons, selling millions in computers there.[12] Best Buy uses it for customer service, which I'll discuss later.

- It's more **diverse** than media: a level field that changes notions of authority. Williams says Twitter "should be as democratic as possible." Ideas should spread not on the basis of fame and followers but on the value of what is said. That's wishful thinking when Charlie Sheen's rants and gags attract more than 4 million followers, garnering more attention than Rusbridger, who has fewer than forty thousand.

- It's an **agent of change**. It doesn't cause revolutions, but it can help spread them.

- It has a **long attention span**. Yes, Rusbridger said that: a *long* attention span. Twitter users "will be ferreting out and aggregating information on the issues that concern them long after the caravan of professional journalists has moved on." Arianna Huffington says that journalists have attention-deficit disorder—they move on when they get bored—but folks online have obsessive-compulsive disorder—once they get their teeth into a target, they won't let go. Case in point: The Nation's Greg Mitchell—@gregmitch—who blogged and tweeted the WikiLeaks story daily for more than 160 days straight.

Williams hasn't made tools to make content so much as he has made tools to create conversations. That, in turn, creates publics. He and Twitter are looking for more ways to help users find the right people and gather around an idea, a joke, a location, an event. Twitter is a serendipity machine. That is the word I hear probably more than any other in laments about the internet's impact on media and the world: that without editors exposing us to new topics of interest, we'll lose serendipity, trapped in echo chambers of predictable sameness. To address the problem, The Guardian played with a serendipity generator to serve arbitrary links. In truth, it produced only randomness. That's not serendipity; it's noise. By my definition, serendipity is unexpected relevance. It is that moment when you say, "Aha! Just what I wanted. Where did that come from?" Through the links I get from people I know, like, and respect on Twitter, I experience many such moments every day. They expose me to ideas, news, and people with a better likelihood of unexpected relevance than a mass-market publication with a single editor could ever deliver. And when I put a thought out there in the Twittersphere, I never know what serendipitous connections will result. "You can bump into people," Williams says. "People putting things out there sparks something."

Twitter users themselves have invented new ways to collect around topics and events. One of their devices is the "hashtag."[13] At a conference, attendees agree to use a shared label—at South by Southwest in 2012, it would be #SXSW12—so anyone can search or click on that hashtag to find out what everybody else is saying about the topic. They also use it to spread jokes. One game I've enjoyed is #boringprequels. Examples from my friend @jimbradysp: *The Pride of the Mets; Dr. Yes; All That Polka;* and *Zorba the Geek.* StockTwits, a company built on Twitter, uses a dollar sign to organize conversations around companies and their stocks. If you want to see what people are saying about Apple stock today, search Twitter for $AAPL. "A unique public is created spontaneously and lasts for a short time," Williams observes. "And then they disperse."

In the end, though, Williams does not believe he created a social network such as Facebook, organized around people and relationships. "We're a public information network," he says. He confesses he didn't know what Twitter was when it started. I still don't think we know what it is. As he recalls, Twitter began by defaulting messages to private because the founders thought users' status updates would be personal. Initially, one status update erased the prior one, so narratives and conversations couldn't build. Just as the users invented the convention of the hashtag, they also invented the convention of the @ reply so conversations could form. (If you want me to see a message, start it with @jeffjarvis, my Twitter handle, and whether or not I follow you, I'll see it in my list of tweets that mention me. I call that my ego search.) Users also invented the retweet or RT convention as a way of passing along interesting thoughts. (When you want to share something I've tweeted with your followers, you hit the retweet button or add "RT" to the start of my tweet and send it on.) Twitter at first resisted the RT idea but eventually acquiesced. Through applications built atop Twitter as a platform, it is adding new capabilities all the time: search, photos, videos. All this functionality is possible because Twitter is a platform that was—and still is—incomplete, evolving in public. Twitter gained wide distribution because outside developers made clients for it, though Twitter later built its own clients, putting it into competition with those developers. The problem is that Twitter remains incomplete in one crucial aspect: It still doesn't have a clear business model; it is protecting its turf because it still doesn't know where its money will come from.

Inside Twitter's collection of chatter lies information of untold value. There's wisdom in that crowd. In 2010, according to the Los Angeles Times, two social computing scientists at HP Labs calculated that analysis of Twitter discussion can be as much as 97.3 percent accurate in predicting a film's performance in its opening weekend.[14] Also in 2010, three computer scientists—Johan Bollen, Huina Mao, and Xiao-Jun Zeng—used a set of keywords to track certain moods in Twitter messages—they fall into the categories calm, alert, sure, vital, kind, happy, and their opposites. Using the results, they found they could predict daily ups and downs in the Dow Jones Industrial Average with up to 87.6 percent accuracy.[15] A hedge fund now uses the formula in partnership with one of the scientists.[16]

In 2005, a remarkable digital artist named Jonathan Harris started collecting snippets of blogs with the phrases "I feel" or "I am feeling." He and his partner, Sep Kamvar, accumulated more than 12 million emotions in four years, displaying the results at WeFeelFine.org and in their book of the same title. Each feeling is displayed as a floating dot. Click on it, and read the emotion. A few at random from the site: "I have a feeling that he's gonna cry and I'm going to be the only one who can stop him because apparently I'm his best friend." "I feel more content, useful, and happy." From the book: "I feel beautiful again." "I feel afraid and think I've lost my way." "I feel good about feeling bad." Harris and Kamvar tracked emotions against location, age, weather, gender, and time of year. You should take heart that feeling "better" (5.74 percent of the emotions tracked) beats feeling "bad" (4.06 percent) and feeling "good" (3.84 percent) beats feeling "guilty" (1.53 percent). As men and women age from their tens to their fifties, they get happier but take a dip in their sixties.

That information comes from listening to people. Twitter is proving to be a tool to collaborate with them as well. Film director Tim Burton started a story with one line on Twitter and challenged his fans to complete it, competing each day to add the next line. The result, frankly, was unimpressive.[17] The crowd does not have the voice and vision of an auteur. But the audience has ideas, and who knows what serendipity its notions may spark in Burton's most serendipitous mind. The tools that Williams made tap into not just the wisdom but also the mood, creativity, news, and experience of the crowd. "More often than not," Williams says, echoing the optimism one hears in this industry, "people are going to do good things and help each other." He helps them do that.

Dennis Crowley: Dodgeball and Foursquare

Visit Dennis Crowley's Flickr photos, and you will find small moments of life: a lunch; a party; his family after Thanksgiving dinner; a video reminding him how to reinstall the window air conditioner next summer; a picture of an odd fruit and a question: "What is this?" (answered quickly by one of his followers).[18] Dennis Crowley is a sharer. He has been for fifteen years, since his freshman year in college, which happens to have coincided with the birth of the web browser in 1994. "Why would you share these pictures of your friends online?" he was often asked. "It just seemed kind of natural," he answered. He doesn't think in terms of privacy or risk. "I'm sharing stuff," he says with a shrug. He uses his blog, Flickr, Tumblr, and Twitter "as a way of lifecasting."

Crowley has taken it one step further. In addition to sharing opinions and photos, he created ways for people using mobile phones to share their locations. While a student at New York University, he and a colleague created Dodgeball, a service aimed at helping young people find one another in bars through texting. Google bought the company in 2005. After two years of frustration working in what was still just a remote New York outpost of the growing California giant, Crowley and his partner quit. Two years later, Google killed Dodgeball—just as Crowley and another partner started a new and more sophisticated location service, Foursquare. Foursquare's smartphone app allows users to check in at a location so they can tell friends where they are. Foursquare added game mechanics, awarding badges and titles for frequent check-ins as a way to encourage use. (Personally, I'm upset that I've never checked in anywhere often enough to become mayor of the place. I go to my local Chipotle so frequently that you'd think I'd at least have been appointed its parks & rec commissioner by now.)

Why share location? The answer, again, is serendipity. Foursquare helps users find people to see and places to go that they might not have found otherwise. The sharing industry has discovered that serendipity is well-served socially. If your friends like something or someone, there's a chance you will, too. "I happen to be a very social person who's part of a very social crew, and I live in a very social city. We're finding that when you opt in to share your location—no matter if it's at a library or a bar or a classroom or the airport—it tends to lead to things that facilitate serendipity," Crowley says. "If you can check in at an airport gate and realize that a friend of yours is three gates down, that makes a connection that wasn't there before. It enables things to happen in a way that you wouldn't be able to do if you weren't participating in these systems."

After he left Google and needed a place to chill, Crowley chose Scandinavia. He took a snapshot of a Google Map with lines connecting the spots he'd visit, and put it on Flickr. In response, he received pages and pages of tips: "It wasn't, 'Oh, you're going to have a great time in Stockholm.' It was, 'Go to Stockholm, find this bookstore, go downstairs, have a coffee, and look at the statues'—very specific recommendations." That inspired some of his thinking behind Foursquare. People will share advice not just for the sake of it but to be generous or influential.

We who live online often end up crowdsourcing our lives, putting ourselves into the hands of others. We go to Twitter or Facebook and ask friends there where we should go to lunch, what we should buy, or how to solve a problem. More often than not, we get suggestions. That's life in future tense, deciding what we will do next. Crowley is fascinated with the present tense, sharing what we're doing *right now*. When we say what we're up to, others can intervene and say, "Oh, you're doing that? You should try this instead." Sharing what you're doing—in public on Twitter or to more controlled groups of friends on Foursquare or Facebook—is "an implicit invitation for other people to give you recommendations or advice." It's also a way to hook up.

Crowley doesn't check in everywhere, all the time. He did check into the Irving Place Starbucks in Manhattan where we met for an interview. "Why not?" he asks. "What's the worst that could happen: someone's going to come up and say hi?" So what's the worst that has happened? "I think I have more check-ins than anyone in the world over a large period of time, like ten years. And the worst thing that's happened is someone came in and sat down at a dinner I was at. It was a friend of mine in the neighborhood. But I was on this date. I didn't know how to tell him to leave, so it was just weird." Thus a new social norm is learned: Don't check in during dates. In general, the worst that happens using services like this—hearken back to Brandeis and Warren's conclusion in 1890—is hurt feelings: "'I was left out of this party,' and the other awkward things that come with a social graph." Crowley says he'd never unfriend someone on Facebook. But he would unfriend an ex-girlfriend on Foursquare. "I don't want to know where she is, and I don't want her to know where I am. So you delete people. Having a social graph that you really have to garden like that is challenging for some people." Friends as flowers. Or weeds.

Some people should not use Foursquare. "If you really are worried about people tracking and stalking you," Crowley says, "there are a lot of tools you probably shouldn't use. If you're a fugitive from the law, there's a good chance you shouldn't be uploading Flickr photos from your vacation." Crowley learned at Google that a service should always offer the means to opt out of a feature and that users should have the option of deleting their data or killing their accounts. The idea of erasing history is controversial online; later, I discuss the European Union's notion of a right to be forgotten. When friends who were about to apply for security clearance or law school would ask Crowley to erase a photo on his personal sites—nothing humiliating; just a beer in hand, that sort of thing—he at first resisted but then would black out the friend's face or change a name to initials. He's not radical about publicness.

Foursquare, like other outposts of the sharing industry, can be seen as more than the sum of its individual members and their single instances of sharing. It also represents the collection of data that comes from all that openness. There's knowledge there—about you, about us, about where we go. Crowley wants to extract value from that data on behalf of users and companies. He wants to tell users how often they've been exercising or flying or nipping a gin at the corner bar. He wants to use the collected data to formulate recommendations. Foursquare's staff, for example, tracks lunchtime check-ins near their Greenwich Village office, sorting them by distance. They then remove any restaurants the employees have already tried, and the most popular new place is where they plan the next team lunch.

I imagine Foursquare working in other ways. As a journalist, I'd like to see who happens to be near a breaking news event so I can ask people what's going on there or have them share pictures. I'd like to be able to use Foursquare in future tense, announcing where I am going to be and then getting friends to join me there (and the clever venue will entice us by offering a free pitcher of beer if we attract enough people). As a local business, I'd like to know more about those faceless people who file into and out of my doors: Where else do they go? What do they like? As a customer of local businesses, I want to point my phone at an establishment and find out who has been there and what people say about it.

When I interviewed Crowley, Foursquare had thirty thousand local merchants using the service to give away discounts or slices of pizza and such. Competition in local is getting fierce. On Google, more than 6 million businesses have claimed their Places pages, and now Google will sell them advertising to drive customers to those pages. Facebook launched Places, another way to check in to addresses. Facebook also launched its Deals feature—coupons for customers—which it gives away free to merchants. The offers make Facebook more valuable to its users, and Facebook can sell merchants advertising packages to get customers to notice their deals. Groupon offers group coupons, and newspapers are trying to copy that model.

Mobile is hot, too. But I predict "mobile" will soon be a meaningless word as we become connected constantly and ubiquitously. iPods and other tablets are transitional devices pointing to a future when we are online all the time and everywhere via our cars and TVs and ambient devices in rooms and . . . oh, hell, just go ahead and implant the chip in my head. Who's to know or care how we're connected? We just are. Mobile will come to mean local and local will mean around me, and that's just another way to say relevant.

Imagine clouds of information around any place, any object, any person. Our annotations of the physical world—made possible by the geographically aware devices we carry—will collect and organize huge pools of information and create incredible local advertising opportunities. That's why Apple, Google, Facebook, and Foursquare are fighting for preeminence in mobile. If you have Google Goggles on your smartphone today, you can take a picture of a bar and grill, then search for what the internet knows about the place. That list could include: What's on the menu? What's the most popular dish? What are the specials? Show me a picture of the decor. Show me pictures of entrées taken by diners (see Foodspotting.com). What's the average bill (see Blippy.com)? Who's playing in the bar tonight? Can I hear her music? Can I buy her songs? Which of my friends has gone here? What do they like? Are there any health-department violations? Can I get a coupon for dinner? Can I get a better deal from the joint across the street? Before I go in, let me tell my friends to meet me there. After I'm done, I may share pictures and reviews and even my bill. Together, this gold mine of information will provide a better basis for a decision than staring at the menu or the logo out front. Today one can also use Layar, an augmented-reality program that shows layers of information atop places as you sweep your smartphone camera across a streetscape. Layar's founder, Claire Boonstra, showed me a vision of the near future when you look through internet-connected eyeglasses to see information about places around you. You might see the schedule for your bus projected onto your bus stop, just for you.

It's not so crazy. A few weeks before Boonstra and I spoke, a company announced ski goggles powered by Google's mobile operating system, Android, that will show you a map of the slopes while navigating them. This is the annotated world.

Crowley imagines that the phone you're carrying will be aware of where you are and what you'd want. "That person you wanted to meet just walked into Starbucks," for example. Or it's lunchtime and your phone knows you haven't checked into a restaurant yet. It lets you know that the place you read about in a magazine two weeks ago is around the corner. Or you just landed in Chicago and here are ten things you should do. By the way, three of your friends just arrived in the city yesterday. Crowley dismisses his first company, Dodgeball, as a "silly, stupid idea." This one, he argues, "can change the world the way Twitter did. It's just a matter of thinking bigger." As an entrepreneur would.

Philip Kaplan: Blippy

If you think it's risky—even insane—to share thoughts, photos, and locations publicly, this start-up will blow your mind. Philip Kaplan cofounded a company that lets you share credit-card purchases and opinions about them with friends and strangers. When I interviewed Kaplan, Blippy had recorded more than 2 million shared purchases—ten thousand a day—representing almost $100 million in total value.[19] Why in God's name would anyone share that?

Let us count the ways. First, people like to be heard. "I talk to a lot of people who already are writing about the stuff they're buying—in particular, Yelp fanatics," Kaplan says, referring to the digital, local, mobile, and social successor to the Yellow Pages that has customers review local businesses, particularly restaurants. What do those users tell him? "It feels so good, I don't know why." "It makes me feel like my opinion matters." "If somebody reads this stuff, it's fulfilling."

Next, people like to show off—to "signal," as Mark Zuckerberg says in the economic parlance of social software.[20] Kaplan recounts a scene from the documentary *Objectified,* in which a BMW executive talks about how people fuss and fret over the car they buy. "They spend forever on this," Kaplan says. "And do you notice when they're driving on the street what fucking color the car is?" If you asked me what shirt Kaplan was wearing, I couldn't answer, though he says, "I'm sure I spent fifteen minutes in the store picking out this shirt." In the end, he contends, "Everybody feels like they have an audience."

There's nothing new in showing off our purchases. Blippy just makes it more convenient to do so by automatically importing transactions from credit cards and online accounts we choose—iTunes, eBay, Netflix, Fandango, Best Buy. Then we check off the items we'd like to share and review. Another start-up by entrepreneur Tara Hunt, Buyosphere, provides a similar service if you forward your email order confirmations.

Kaplan and Hunt have identified a social benefit to this kind of sharing. Kaplan tells the story of one Blippy user, Matt Cutts, the Google engineer charged with outsmarting and exterminating spammers. He bought a networked scale that sends your weight to your computer—so you can track and chart your progress—or even to Twitter so friends there can keep you honest. (The New York Times' Brian Stelter used such a scale for his Twitter diet). As soon as Cutts bought it, Blippy saw another hundred users order one. They had discovered the product through Cutts and trusted his judgment. That is social shopping, a force far more powerful than any banner ad on a web page. That is what Facebook's Zuckerberg wanted to create with his Beacon ad service. On Facebook, as Zuckerberg acknowledges, making users' purchases public sneaked up on them. But Blippy is blunt. You are there only to share what you buy.

Perhaps Blippy's greatest value as a tool could be to create transparent markets, revealing the prices people pay. Transparency inevitably benefits the buyer. "We can absolutely show you, for example, the cheapest HDMI cable that has ever been bought by anyone, pretty much anywhere . . . and it costs fourteen cents," Kaplan says. "If everybody knew what everybody else was buying and how much they were spending, well, nobody could ever get ripped off." His service can gather more data of interest: price fluctuations, discriminatory pricing, economic indicators, and trending products and stores. Take such data about shopping out of the hands of the retailers, put the information into the hands of the collected shoppers, and—as with other once-secret data made public—power passes to the customers.

Blippy is not the first or only example of such collective power. TripAdvisor is a small miracle. I have found informative and moderately accurate reviews by fellow travelers even of obscure lodges in remote towns in South Africa. The company found that when it allowed travelers to share not just with the world but more immediately with friends in a Facebook app, the number of reviews it received exploded. Today TripAdvisor has more than 40 million reviews of more than 1 million businesses—450,000 hotels, 650,000 restaurants, 125,000 attractions—in 80,000 cities with 6 million photos.[21] To provide this information, users violate their own privacy to the extent that they share their travels and experiences. Why do they do it? To be generous? To empower fellow consumers? To show off? To create? I don't know why, but I'd love to see research on the question.

Blippy began by sharing every purchase users made on a registered credit card. Sensibly, it shifted strategy and allowed users to decide whether to share each purchase individually. That change didn't come from privacy concerns—"We didn't even think about privacy stuff in the beginning," Kaplan says. Instead, Blippy found that when users choose which purchases to share, they are more likely to write reviews of what they've bought. That produces more information. Blippy pivoted to emphasize reviews over purchases. In the future, Blippy could issue its own credit card—a social credit card—to give it more data. Soon you'll be able to simply wave your Android phone at a cash register to pay. Your phone will know who you are, where you are, what you're buying, when, and for what price. And your friends can know, too, if that's what you want.

Kaplan has bigger plans. He wants to disrupt retail and "commoditize every store in the world" to give his users better bargains—and to make more money. That disruption of retail has begun. Two killing pressures already threaten to crush stores: transparency of pricing, which squeezes profits, and the efficiency online competitors gain by consolidating merchandise in a few locations, which saves operating costs and capital for inventory and real estate. Blippy—among others—could disintermediate retailers, both online and off, by consolidating data about merchandise, prices, and service and then offering the opportunity to purchase via Blippy. You see your friend buy that Wi-Fi scale, you read real customers' reviews of it, you want it, and Blippy finds the best price and gets it to you. Store? Who cares what store it comes from? Stores and catalogues have long been curators that select merchandise, stock it, and sell it. Your friends can now be your curators and merchandisers. You need not care who fulfills the order.

Are there limits to the purchases people will share? Blippy still hasn't proved that most people want to share purchases at all. Would you share the cost of health care? A student in my entrepreneurial journalism class, Jeanne Pinder, launched ClearHealthCosts to bring transparency to health-care procedures. Her first good example of insanely inconsistent pricing is the colonoscopy. In her data, the procedure costs anywhere from $500 to $3,500 in the New York area. Would you share the cost of porn, then? Isn't that the limit? "People always give the porn store example," Kaplan says. "Just to make a point, I did share a bunch of stuff like that. The most embarrassing store I can think of is a store in the Castro [in San Francisco]. It's the gayest neighborhood in the world and the gayest sex shop. The name of the store is My Mother Doesn't Know. . . . I was like, 'This would be great on my Blippy.' So I went into that store and I bought something—I won't tell you what I bought—just to make a point. And the point is, nobody gave a shit. It actually wasn't that interesting. Everybody's like, 'Oh, yeah, you went to a sex store. Woo-hoo.'"

"People think their lives are a lot more embarrassing than they are," Kaplan says. He figures that perhaps 1 percent of people go to porn stores (though he also points out that after the internet, nobody needs to anymore). "Those 1 percent know who they are." So they're not going to use Blippy. Neither are the 1 percent who are cheating on their wives, buying presents for mistresses. "The other 99 percent of you have nothing to worry about," Kaplan argues. "But what I will say is, 100 percent of people have embarrassing thoughts. . . . That's probably why everybody's freaking out about privacy."

Kaplan is not a private man. He's another sharer. Back in the technology industry's hairpin boom-and-bust of 2000, he wrote a newsletter chronicling its excesses and idiocies, which became the web site and book FuckedCompany.com. He also wrote about himself and—since Howard Stern is an idol—"I used to make masturbation jokes. And then I would talk about my private relationships and girls I was dating. A lot of people hadn't really ever seen anybody on the internet be so personal." Well, that soon changed. What's changing is our accepted norms. Back in the day, when we all took our pictures to the photo store, it weirded us out that some stranger was looking at our family scenes. "You don't get any benefit from one guy who works at a photo store looking at your pictures, but you might get a benefit from a million people looking at your pictures. But the tradition is, 'Nobody saw my photos except possibly that one guy, and that freaks me out. Therefore, I don't want anybody to see my photos.' There just wasn't the technology to do it."

In *What Would Google Do?*, I told the story of Flickr's accidental decision to default photos to public. Other photo services made the logical assumption that if we put pictures online, we'd sure as hell want them to be private, right? Not Flickr. When it let strangers share and comment on photos, magic ensued: Search for "funny" photos, and you'll be sure to be amused ("sexy" is more a matter for the eye of the beholder). Communities form around interests, events, artists, and cameras. Flickr created algorithms to use such data to surface interesting photos.[22]

As we talk about privacy, Kaplan launches into a rant about the so-called Ground Zero mosque in New York and how the media whipped up a frenzy over it. "This is one of those things that is completely manufactured by the press," he contends, arguing that the frenzy about privacy is similarly engineered. "I think the reason why so many people freak out about privacy and they yell about privacy is because they've been told to. Somebody said, 'This is offensive and you should pay attention.' It totally holds people back from what is potentially a very fulfilling experience." He returns to this notion often. "It's not that I'm against privacy. I think privacy is awesome. I just think fewer things need to be private. . . . Is there anything bad that will come from somebody knowing that you went to Starbucks this morning?" And bought a grande, decaf, nonfat, no-foam latte and paid $3.80 for it?

Josh Harris: *We Live in Public* and Wired City

How far does all this go? From sharing our photos to thoughts to locations to purchases . . . to where? To Josh Harris, that's where. He is the artist, philosopher, and entrepreneur of publicness, our public savant.

Harris is best known as the creator of two radically public art/publicity events captured in the documentary *We Live in Public*. In the first event, 1999's "Quiet: We Live in Public," he built what a Wikipedia writer calls a human terrarium[23] and others liken to the Panopticon, an eighteenth-century vision of a prison whose inmates never knew when they were being watched because they could be watched all the time. In Harris' experiment, a hundred volunteers lived under the constant gaze of cameras that broadcast their every moment to the web and the world, making the very private—showers and toilets, sex and fights—very public. Harris was Warhol 2.0, stopping the clock on fame. These were the all-sky, no-clouds days of the early web—before the crashes of the dot-com stock market and of Osama bin Laden's jets into the World Trade Center. There seemed to be no limits. So Harris pushed the question to its limit: Just how public do you want to be, people? The effort devolved into a perhaps predictable bacchanal of discord and decadence, an outbreak of attention excess disorder. New York police shut the project down on—oh, how metaphorically appropriate—New Year's Day 2000, Y2K. Did Harris discover the limits of publicness, or was the experiment rigged to explode with the people he picked, the rules he set, and the heat of the spotlight? In any case, we got a new illustration of the extremes of publicness.

Harris continued in a new laboratory, wiring his apartment for video—thirty-two cameras pointing everywhere—and for sound, to capture the discussion of anything he and his girlfriend, Tanya Corrin, did. Corrin says she was his girlfriend. Harris says that she wasn't and that he cast her in the role for his reality show (but if she was cast, was it reality? . . . the mirrors in this hall are wobbly). Harris wrote an algorithm that decided which camera to display to the net. Projectors around the loft let the couple interact with what the world was saying about them in a constant chat. Twice a day, they'd open the phones and "converse with our watchers."

The critical moment in *We Live in Public* came after Harris and Corrin fought—he got too physical and frightened her, and she banished him to the couch. Harris says that was out of character—not in her personality, her inner script. He says the spark came from "her people," the public who talked with her online. "They gave her the strength, but there was a trade-off. The trade-off was, they took a little piece of her individuality." That is, he argues, she let them make her decision for her. She crowdsourced her life. Or, as he puts it, "A part of her brain was augmented by beings from the ether." A conversation with Harris includes moments like that, when metaphors leap to beings from the ether or the Mayan calendar and its prophecy of apocalypse in 2012—and you may wonder whether you've just traveled through a wormhole to Area 51.

But Harris will return to reality if not practicality. He is also an entrepreneur. He started and sold Jupiter Research. He founded Pseudo.com, probably the first TV company for the web age. Too bad he was so ahead of his time and started it a decade before the web was capable of serving video well. It went bankrupt in the bubble of 2000. He owned and ran an apple orchard in upstate New York and started another video venture. Now he is trying to start another company, Wired City.

At Wired City, Harris wants to take a soundstage and make it home and work for two hundred people. The problem at the *We Live in Public* project, he says, was that it took over its participants' lives just eight hours a day. People had jobs and activities outside. In Wired City, being part of the game is your work, sponsored by brands. It is not a reality show. It is a massive, multiplayer game set in reality. "The difference between what I'm talking about and World of Warcraft is that you're playing with people's lives, for real."

The game is played from home studios, where contestants vie to win a spot in the Wired City studio. Harris envisions ten thousand players

holding up a tube of Crest at the same time—"simultaneity matters"—
and if Crest doesn't sponsor the moment, they'll hold up Colgate the
next day. In this future of marketing—"the next golden age of the advertising business"—he imagines Crest having representatives. At level one,
the Crest rep helps with your flossing; at level ten, she's an oral surgeon.
He imagines Crest supplying users with oral hygiene monitors. (That's
not far out. I just saw a toothpaste commercial promoting such a thing.)
He sees using cameras to connect twenty people brushing their teeth
simultaneously. If they play it right, they're rewarded with Derek Jeter or
Cindy Crawford joining in from their own bathrooms. "The whole idea
is to elevate that microday-part into an engaging, entertaining experience." By making your tooth brushing public, you are rewarded. And
Crest? It doesn't want banner ads or billboards. It doesn't want to control
the so-called desktop of your computer. "Crest will gain control of the
bath top." So now each "microday-part"—standing at the stove, sitting
at the makeup mirror, driving—is made social and public, part of a game
with rewards and sponsors.

Harris tells me he wants to wire a mall. "When you go into a mall
now, you're like a zombie. You have no expectations of having any meaningful social interaction." True. You're surrounded by strangers. You shop
for a washing machine, and the sales guy "has a very, very low likelihood
of knowing exact shit about the machine." Imagine if all our connected-
ness gave us access to the expert at the factory and fellow customers who
know and use the product. I know two commercial visionaries who have
similar plans. (One of them, Shawn Samson, is working on a project to
turn retail and marketing on their heads.[24] I'll look at the other—Best
Buy—in the next chapter). Harris imagines that the store and the manufacturer will put you together with twenty other people shopping for the
same device. That group becomes an instant social network. It's a lite and
temporary network—you're not going to invite one another to your wed-
dings. But the group can make use of its common interest in this device.
You have people to call on if you have a question or problem, just as
people do online today around computers or phones. "They can use the
machine as a relationship focal point," Harris says. "You've connected in

time and space in a way that you could never have done before"—around a product. Isn't that marketing nirvana? "In the mall of the future," Harris predicts, "they won't let you in unless they know who you are. And when you get there, you're meeting people you know." He happens to be describing Senate Commerce Committee Chairman Jay Rockefeller's nightmare. At a hearing on privacy, Rockefeller testified:

> Imagine this scenario. You're in a shopping mall. And while you're there, there's a machine recording every store you enter and every product you look at, and every product that you buy. You go into a bookstore. The machine records every book you purchase or peruse.
>
> Then you go to a drugstore. The machine is watching you there, meticulously recording every product you pick up, from the shampoo to the allergy medicine, to your personal prescription.
>
> The machine records your every move that day, every single move. Then, based on what you look at, where you shop, what you buy, it builds a personality profile on you. It predicts what you may want in the future, and starts sending you coupons. Further, it tells businesses what a good potential client you may be and shares your personality profile with them. . . .
>
> So, this sounds fantastic, something like out of science fiction. But this fantastic scenario is essentially what happens every second of every day to anyone who uses the Internet.[25]

So? Companies already know a lot about your purchases. Your credit card knows where you shop, your grocery and drugstore cards store what you buy. Marketing databases have collected and sold such information for years. Today, I can go to a database sold by Acxiom and buy a list of names and addresses of, say, single women in their thirties with high income who've finished high school within 1 mile of an address. Sound creepy? That has been happening long before the internet. That is how junk mailers get your address. What's the harm of online targeting by comparison? What's the benefit? That is a calculation for us to make. I would like to see independent networks of people I know built around products, services, and brands I'm interested in so I can get credible advice and help. That would be far more useful to me than commercials blaring and repeating sponsors' messages. So I am willing to announce myself as an iPhone owner or camera shopper. No problem.

Given these considerations, Harris' ideas don't look so far out. He envisions having "a meaningful social interaction" in his wired space instead of "going to a mall and being a zombie walking around." His Wired City sounds more human and less cold and frightening than the mall or a marketing universe that wants to, in the argot of the field, "capture your eyeballs." Harris doesn't see this progression as good or bad but inevitable. "The audience are going to demand self-surveillance," he predicts. It has already begun.

The Radically Public Company

Imagine

Over the next five years, Mark Zuckerberg says, most industries and many companies will be reimagined and redesigned as social enterprises.[1] In this chapter, I will first imagine the radically public company. Then I will look at examples of companies that are succeeding and failing at becoming more public.

The radically public company would encourage all its employees to use the tools of the public net to **have direct and open relationships** with customers—answering questions, hearing and implementing ideas, solving problems, and improving products. The clearest lesson of the social web is that people want relationships with people, not with brands, spokesmen, rules, robots, voice mails, machines, or algorithms. Social tools permit even big companies to return to the days of doing business over the cracker barrel, eye to eye, with familiarity and knowledge of each customer and her needs. Direct connections help customers trust a company's people and build real relationships. I have this kind of relationship with my corner dry cleaner when I let them know they're massacring my buttons so they can fix the problem. Shouldn't I have such a relationship with any company, no matter its size?

The radically public company would **open up as much data as possible** about its products and process, including even design specifications, sales and repair data, and customer feedback as well as the provenance of the ingredients and parts it uses. Insane? A company must decide whether greater value lies in its secrets or its relationships. It needs to calculate what benefits might accrue from transparency. Are you likely to trust the companies that fought New York legislation requiring them to disclose ingredients in their cleaning products?[2] Or will you wonder and worry about what they're hiding from you? When Bank of America reportedly opened a war room to strategize its defense against a laptop full of secrets WikiLeaks was threatening to dump, it occurred to me that every company should open such a room and ask what it would be ashamed to have exposed. Then, of course, it should stop doing those things.

The radically public company would **become collaborative,** opening up design, support, marketing, even strategy to its public, releasing plans and beta products in progress. That doesn't mean it would become a democratic enterprise, run by a committee of the whole world. It means opening up so the right people can have a say at the right point in the process of product creation so a company can hear and implement the best ideas. The company's role is to feed and lead that process. As you'll soon see, it's possible to design even cars this way.

The radically public company might—just might—be able to all but **eliminate advertising**, relying on customers to sell products for them. Wouldn't that be nice: the company that doesn't want to bother us?

The radically public company would, in clear language, **reveal and explain everything it does with customer information**, giving customers a simple means to opt in and out and to correct data.

The radically public company would make all my **data portable**, letting me leave and take my information—my emails, purchases, preferences, connections, creations, friends, everything—elsewhere. The radically public company would be confident enough to do that. Google has a department devoted to such portability. They call it the "data liberation front."

The radically public company would **open its books**, even its salaries, to public view. A step too far? Probably. But consider how giving employees information about the bottom line could help them pull together toward common goals and understand the impact of their work. No longer would the grass be greener on the other side of the cubicle. Consider how such transparency could take the poison of secrets and politics out of an organization. Consider how it would enforce equality and fairness.

The radically public company would support and operate under **open standards**. That way, it could run more efficiently, using off-the-shelf parts and software, benefiting from others' innovations. Even secretive Apple released its WebKit browser engine as open source, seeing the benefits in creating a standard and allowing others to contribute to it. WordPress, a leading blogging platform, made its software open source at WordPress.org and then built a commercial blog-hosting company, WordPress.com, atop it. Anyone else can use the same code to build a competitor to WordPress.com—and if they do, WordPress.com benefits every time the competitor improves the code.

The radically public company would see itself as a **member of an ecosystem** more than as a conglomerate that wants to control all it surveys. Such a company would understand the value of every relationship, even relationships with competitors, to create both value and efficiency. The radically public company might also see itself as a **platform or network** more than as the owner of assets. It would foster others' success. YouTube is trying to become a newfangled network, not by creating and controlling content but by distributing it via its own site and millions of others and by selling advertising to support the best creators. That could make YouTube scale to be bigger than any old network.

The radically public company would institute **new kinds of governance**. What if it had a Constitution and a Bill of Rights that everyone—employees, customers, suppliers, and executives—could rely on? What if it had a Congress of Collaborators and a Senate of Employees who could propose laws, products, and procedures to the chief executive, who would still be in charge but would have new ways to get advice? If that all seems just too new, then start here: What if a company had a Court of Customers that could adjudicate disputes, absolving especially market-controlling companies from accusations of monopoly control? Google should have such a body to settle appeals when it rules that an advertiser is a spammer. Now Google holds the power of God over these firms. Such exercises of power could lead to greater regulation. Google would be wise to share this power with constituents—users, publishers, advertisers—that share Google's interest in eliminating spam.

Finally, the radically public **CEO would become the leader** of something more than just a company: a community? a movement? a mission? Yes, a company must be profitable to be sustainable. It must build its shareholders' value. It must win against competitors. None of that changes. The question is, can a company be more than a company? Must it? University of Virginia professor Siva Vaidhyanathan, author of *The Googlization of Everything,* has debated this point with me on Twitter and YouTube: Can a company be anything other than evil, to use Google's loaded and perhaps shallow term?[3] Should it try to do anything other than increase its stock price? Does its obligation to maximize profit obviate the opportunity to have a higher mission and to operate under new ethics and norms? Economist Umair Haque argues that the cost of companies doing evil has risen dramatically in an era when customers can call them on their sins. When companies are forced to operate in the open, are they also forced to become better companies?

No company I know of is doing all I suggest above. Likely none ever will. But companies are trying bits of everything on this list. Here are some examples:

Open-and-Shut Case Studies
Manufacturing: Making the Homer

Local Motors is the flea that wants to bite Detroit on the butt: the new, open, collaborative, small, nimble, efficient, and cool car company. Founder Jay Rogers' vision is a string of automobile microfactories, forty-plus employees each, across the United States. Each will manufacture a few designs that have been created with the help of customers who will buy and even help build the cars. The facilities—he opened his first in Phoenix in 2010—act as showroom, factory, repair shop, school, and community center, where customers can hang out over burgers and beer once a month. Online, they hang out at Local-Motors.com, where they help design the cars.

Rogers argues that though people say they care about what's under the hood, they care more about a car's look and feel. He saves money and much time by using mostly available parts—the first model has a BMW engine, a Ford F-150 axle, a fuel door from the Mitsubishi Eclipse—and he and the community put their effort into the design around that base. With those economics, Rogers says, he can make a profit after selling just two hundred to three hundred cars in a line, and he promises to build no more than two thousand of a model.

After I wrote about the idea of collaborative car design in my last book and before I met Rogers, critics hooted that the product I envisioned would turn into the Homer. Designed by Homer Simpson with shag carpeting, giant cup holders, and two bubble domes (one for quarrelling kids with optional restraints and muzzles), the Homer ran its manufacturer into bankruptcy.[4] "Jeff Jarvis may be a really smart guy on a lot of subjects, but the auto industry apparently isn't one of them," snorted Autoblog.[5] The idea of releasing cars as products in process was "ridiculous on its face," the blog said. And it argued that car companies already rely on customers for design: "They do that with the dollars they spend on cars." Cars are bland, Autoblog contended, because that's what people want. Really?

Here's the test. Local Motors powers what it calls "cocreation" of cars. It's not democratic design. It's more of a republic. Rogers, as CEO, remains responsible for building economically viable products that meet government safety standards. He doesn't hand over decisions to simple voting. The company uses an algorithm—which is not fully revealed, to guard against gaming—to give greater weight to the opinions of people buying the cars or contributing to the design process. He respects his community and defers to its judgment whenever possible. If he ignores the community's desires, they will use Local Motors' own tools to revolt (just as Facebook members use Facebook to organize protests against changes they don't like).

When Rogers opened Local Motors, the company's first model, he said, would be an off-road muscle car. The community he attracted was thus self-selecting. Greenies who wanted a little electric car would not come to fight over design decisions. They would have to wait for their own model and community. Designers submitted proposals (to date,

Local Motors has attracted an amazing forty thousand designs). Korean-born Sangho Kim posted his sketch, inspired by the P-51 Mustang fighter plane, in 2008, two years before he graduated from design school in California and two years before the first car was built. "Community feedback was immediately positive," says the Local Motors web site, "and many demanded the car be built." Kim won $20,000 in prize money and, working with the company's staff and the community—the site acknowledges 165 fellow users as codesigners—he saw his car, the Rally Fighter, come to life.

Here's my favorite part of the story: In the process of completing the design, the community loved an original concept for a taillight lens. OK, Rogers told them, but he priced out what it would cost to tool up and make the lens. It would have added $1,000 to the $50,000 price of each car. "Never mind," the community responded as one. They hunted through other options and settled on a $75 Honda part. From the look of the finished car, I would never know that the lens had such humble origins.

Get that: Given the opportunity, the respect, and the necessary information, the community of customers made both design and economic decisions. Detroit goes to comic lengths to keep its designs secret, under actual wraps, away from media and drivers. When the cars are revealed, they're often dull and rarely distinctive. By the time customers see and sit in the cars and could make suggestions, it's too late to listen. It would take years to incorporate their ideas and desires: More cup holders! . . . A plug for my iPod! . . . Aerodynamics that won't bust an eardrum when you open only one window! . . .

Local Motors does the opposite. It doesn't just open up the process, it open-sources the designs, licensing them under the copyright regime known as Creative Commons, which allows the creator to set clear and simple conditions for reuse. Publishing complete and detailed specifications allows community members to design component parts. It also facilitates an aftermarket of "modders" to make add-ons they can sell to car buyers through Local Motors. Local Motors' auto, like Apple's iPhone, thus becomes a platform for other entrepreneurs. Even Rogers' team would prefer to keep the data secret. Rogers fights that reflex, arguing,

"The benefit of keeping it back does not outweigh the harm of being like everyone else—if we take the bold step of being the first people to collaborate." That is what will make his company famous, he believes. His secret sauce is publicness.

Compare and contrast Local Motors with Toyota. Sudden acceleration problems, which the company was slow to acknowledge or repair, led to a reported 89 deaths and more than 330 related lawsuits in the United States alone as of mid-2010.[6] Toyota's reaction to this disaster should have been radical transparency. I say that, as a public company in the true sense of the word, Toyota should publish the entire database of its repairs, reported problems, and customer complaints. If that information had been public in the case of the sudden acceleration problem, someone—inside Toyota or out—might have spotted the pattern and discovered the problem long before eighty-nine people died and Toyota ended up with a moral and brand catastrophe. After such a calamity, it takes bold action to regain the public's confidence.

What Toyota gave us instead were Sunday-morning commercials assuring us that our safety was one—yes, it said just one—of the company's highest priorities. My family, parents included, owns four Toyotas today. Over time, we've bought as many as eight. They have been amazingly reliable, even as I've blown past the deadlines for scheduled service. I like driving them. I had every reason to keep buying them. But now I may not. The company violated my faith. Toyota thought it was making cars. It wasn't. It was making trust. I placed the lives of my family in those cars. If I cannot count on Toyota to keep them safe, to care about their security, to make that *the* highest priority, I can no longer do business with them.

I ask Rogers whether he would reveal repair records. His instinct: Why not? By using the open community to catch and fix problems earlier, Rogers expects a payback in lower insurance rates. He has some concerns about privacy: If Mr. Jones keeps backing into poles, is that because Mr. Jones is a bad driver or because there's a problem with the placement of the rearview mirror? It is through community discussion of such questions that Rogers hopes to improve the company's cars. "Everything we do is beta," he says.

A beta car? That sounds absurd, I know—worse even than the Homer. Who wants to drive an unfinished vehicle that could—in both the software sense and the automotive sense—crash? I'm not suggesting that Local Motors or Toyota should produce anything less than the best possible car. But Rogers is more realistic than the auto industry. He acknowledges that a car can't be perfect; it can always be improved. The process of making the product should continue, openly, after the car leaves the factory and showroom. At Local Motors, the privilege of buying the first Rally Fighters also carries a burden, one that early adopters of other gadgets know—that they may suffer more defects than later buyers. At Local Motors, being an early adopter also carries a responsibility: The first customers help improve the model for later customers. Local Motors customers don't just buy the car. They buy the process.

Toyota's warranty should be more than insurance to fix my car when it breaks. It should offer me a subscription to a stream of updates and improvements. Toyota could always make its cars better. It costs money to fix problems that may never turn into disasters. But a disaster could cost much more in money, reputation, lives, trust, and brand value than a lot of fixes and improvements along the way. So this practice would be a warranty for the company as much as it is for the customer.

As computers and software control more and more aspects of cars, improvements can be distributed like songs to iPods through the car's internet connection. Cars will be online, as most every device—TV, refrigerator, washer, furnace, security system—gets hooked up to the net to gather data, make usage more efficient, and deliver improvements. All those devices should have APIs—application programming interfaces, or the sets of instructions that enable independent developers to create programs for them. Imagine going to Food TV's web site and getting not just recipes but programs for your oven, with the perfect temperature variations to roast your turkey or bake your soufflé. I've seen a design-school project from Japan that proposes just such functionality for a next-generation stove. By opening the API, the manufacturer encourages users to become, in Local Motors' word, cocreators in a process of design and improvement that never ends. Which oven would you rather buy:

the one that stays the same from the day you bought it or the one that evolves? The one controlled by the company that made it or the one that opens up some control to its community of customers and collaborators? And which company would you rather be: the one that has no relationship with its customers—unless they're angry—or the one that works with its customers to make better products?

Local Motors believes it is a part of a movement toward what it calls cocreation. So it played host to a conference of cocreationists at its Phoenix factory in 2011 with talks scheduled by:

- MESH01, a cocreation platform that allows shoe and clothing manufacturers to open competitions among designers.

- Quirky, another platform that brings products to market through collaboration among customers and designers (it has a bunch of cool cord organizers and kitchen utensils).

- Wired magazine editor-in-chief Chris Anderson on the "maker movement" of "small-batch manufacturing" using such innovations as 3-D printers.

- LEGO on consumer-generated products.

- An ad agency on collaborative marketing.

- Huggies on its Mominspired entrepreneurial program, investing in customer-created products from a buzzing, cooling, pain-relieving gadget to a sippy cup that doesn't force kids to do backflips to drink.

- Hallmark, which has held twenty-five contests attracting thirty-five thousand submissions for six hundred cards made by the public.

Consumers aren't just consumers anymore when they become cocreators.

Technology: Naked geeks

Google: open or closed? Both. You can't get into Google's offices to so much as have lunch with a friend without signing a nondisclosure agreement, a presumptuous policy. Closed. Then again, Google releases products in beta so it can hear from users how these new services should be finished, which is a call for collaboration. Open. Google won't reveal basic statistics about its size: how many servers it has, how many pages it monitors. Closed. For years, it also refused to tell its media partners that distribute the ads Google sells what share of revenue they were getting. Closed. But after pressure from larger publishers it relented and revealed its cut.[7] Open. It uses proprietary software for much of its infrastructure. Closed. But it released its Chrome browser and operating system and its Android mobile phone operating system as open source.[8] Open.

Google has the reputation of a fortress, but Eric Schmidt argues that the company is transparent. "People confuse transparency with naked-ness," he says. Corporations have proprietary information. They can't discuss business results casually. "Often people will say, 'If you're a leader in transparency, then why don't you publish all your source code?' There are a number of reasons why we don't do that. One has to do with the algorithms getting gamed. Another has to do with cybersecurity," Schmidt says. "But hopefully most of the rest of what we do is very transparent. I don't know how we could be more transparent." His prescription for transparency is to write down your policy. When the public had questions about some decisions regarding Google's home page, Schmidt told staff to write down what they do. When they do, they're held to it. "The credibility of you as a firm will be determined by whether there's a gap there or not."

Google, I believe, is opportunistically open. At least it recognizes the opportunity. By releasing its Android operating system for free to manufacturers and phone carriers, Google has used openness to challenge the supposedly invincible iPhone and iPad. Schmidt said in a meeting with editors in Davos in 2010 that the winner in mobile will be the company whose operating system has more users. That will attract more developers, who will create more applications, thus adding more utility and attracting more customers.[9] A year later, Schmidt told the Harvard Business Review that Google's 2011 strategic initiatives were "all about mobile."[10]

Openness, though, is in the eye of the beholder. In Germany, at an event on privacy organized by the Green party, I joined a panel that included Thilo Weichert, privacy commissioner for the state of Schleswig-Holstein. He sputtered, red-faced, at the mention of Google. "As long as Germans are stupid enough to use this search engine," he said of his own constituents, "they don't deserve better." He dismissed Google's claims to virtue. "Google is the worst example of openness, transparency, and the willingness to meet the democratic needs of our society," he proclaimed. He went so far as to compare Google to China and Iran. How? Because they "surveil without the surveillant being surveilled." That is the core of a complaint I hear from Weichert and other critics: an imbalance of transparency. Google knows more about us than we know about it. "This data is being exploited by Google for commercial purposes and not for democratic ones," Weichert argues. Well, yes. It is a company, not a government agency. Weichert resents its size and success. The lesson, I think, is this: The greater a company's control of a market, the more trust it must earn, and one way to earn it is through transparency.

Cable and telephone companies are practically dictators in their markets. We resent them for it. They force us to pay for channels we don't watch. They make us wait all day at home to do business with them. They nickel-and-dime-and-dollar us to death. They threaten to decide what we can and can't do on the internet that we pay them to bring to us. But now we have a weapon to fight back. We have Twitter.

Comcast, the largest cable company in America, learned the hard way that its imprisoned customers could start mobs. Customers used Twitter to complain about outages, bad service, and rude employees. Then in their midst appeared Frank Eliason, a corporate vice president who tweeted under the name @comcastcares. He solved problems. He answered questions. Sometimes, he gave customers answers they might not have wanted to hear, but he built credibility that way. He was candid about Comcast's problems, with a rare sense of corporate humor. I watched him at a Salesforce.com event when he came onstage and said, "Customer service . . . We're well-known for service, aren't we? . . . C'mon." Pause for laugh. "We're actually working very hard to improve the customers' service." After he left Comcast for a bank job in New York, Eliason told me via email that "we are now able to know so much more about our customer. They are no longer a number, group, classification. Instead, they are human beings with specific needs." And through Eliason, customers also saw his company as more human.

One day on Twitter, I saw Robert Scoble, a blogger, tweeter (with 184,000 followers), and video maker, lamenting that his cable connection had gone down just as he was to make a presentation. He tweeted about his crisis to @comcastcares, knowing Eliason would see it—and see to it. Sure enough, Eliason replied in moments. Scoble himself was a prototype for @comcastcares when, in 2003, he blogged for Microsoft with disarming frankness, especially for the time. Scoble dared to tell Microsoft's imposing CEO, Steve Ballmer, at a public event that the company needed a more human face. Microsoft hired him to become that face. Not only did he promote and defend his employer, he also criticized and complained about it. He praised competitors. That openness earned him credibility that rubbed off on Microsoft.

"Let's say you're a lawyer," Scoble advises. "In the old world, you were recognized as a lawyer by the books you had in your office and by the degrees you have acquired. But in the modern world, we can't see that. Instead, we go to Google and ask, 'injury lawyer San Francisco.' Who comes up highest? Not the guy with the most degrees but the one who has had the most people link to him or her. How does that happen? Well, the lawyer who blogs and shares his or her information gets linked to more than the lawyer who doesn't, even if the lawyer who doesn't has more stored-up knowledge and a degree from a better school." Pause for ironic emphasis. "The internet is so unfair!" Scoble wants to hire lawyers or plumbers who demonstrate that they know what they're talking about. "So the lawyer who teaches me how to do my own law will get the business before a lawyer who doesn't," he says. "I can look at what they actually do and decide if it's right for me. Hoarding information isn't smart anymore." Sharing information cannot be delegated to a PR firm or a marketing department. Everyone who knows anything that customers care about should be communicating with the public.

At Twitter, the company eats its own dog food. If its employees didn't use Twitter, that wouldn't look good. So they do. You can read them all at twitter.com/twitter/team. Naturally they tweet about their work. What are the rules? They're simple and obvious: Don't reveal secrets. Don't be an idiot. There's one other, more subtle guideline, Evan Williams says: Because Twitter is under a bright spotlight with fans and tech blogs scrutinizing its actions, managers caution employees that a simple tweet can be blown out of context. If an employee says, "Whew, worked all night on our big release!" the geekvine may suppose there's some revolutionary new product cooking. Instead, this guy may just be working hard on a new bill-paying program for the accounting department.

I ask Mark Zuckerberg, too, about starting and running Facebook so much in public sight. What is his advice for other companies? "Transparency increases integrity," he says. "In the strictest definition of the word, integrity is basically saying one thing to everyone. That's true for people, and I think it's true for companies, too." As I said earlier, there is some debate whether people should try to maintain multiple identities—for work, home, school—or one. Zuckerberg says that companies can afford only one. They do have different audiences—users and advertisers, in his world—and "it's really easy to say different things to them. But in a more transparent world, you can't."

Media: Just another brick from the wall

John Paton heads Journal Register Company, a poor, neglected newspaper chain milked dry and left to bankruptcy before he took over as CEO in 2010 and brought a fresh wind of publicness. Paton blogs with surprising candor about the state of the company.[11] He shares the financial performance of his private corporation—with a broad brush—not just with employees but also with the public. He decreed that his papers must put digital first, print last. Reporters share stories in progress online and involve their communities in reporting, making their process public. They are building their digital future rather than protecting their print past. In his so-called Ben Franklin Project, Paton challenged his staffs to find ways to publish their papers using only free and open web tools— saving Journal Register more than $12 million in technology costs. They shared lessons from Ben Franklin with the industry, including outside developers who volunteered to help solve problems. His Torrington, Connecticut, newsroom moved out of a rotting, claustrophobic building that would have given Charles Dickens shivers and into an open space where the public is invited in to attend news meetings, meet reporters, take classes, have coffee, and hang out. He has webcast once-private meetings of his editors, publishers, and advisers (I am one of them). At a meeting of newspaper publishers in 2011, Paton—@jxpaton—delivered his speech in tweets, the most incendiary of which advised, "stop listening to print people and put digital people in charge—of everything."[12] None of this innovation, in itself, will make Journal Register's business work. That still depends on the known dynamics of serving the community with journalism and selling ads. There's a way to go before the company reaches Paton's goal of 50 percent digital earnings by 2015. Still, after a year of digital first, online revenue is growing fast enough to nearly replace lost print dollars and the company is profitable.

Alan Rusbridger, editor-in-chief of The Guardian in London, talks about the "mutualization of news."[13] In 2010, a court injunction gagged his newspaper from reporting on a parliamentary question regarding the dumping of toxic chemicals in Africa related to the company Trafigura. In frustration, Rusbridger went to Twitter to tell the public that he was not allowed to tell them what was happening. He was confident what would result: In forty-two minutes, the readers dug up the subject of the gag on their own.[14] They tweeted it. The news spread faster than a polluter's toxic ooze as "Trafigura" became a trending topic on Twitter. The injunction became pointless, and the company and law firm that had sought it learned a lesson about the power of publicness. Says Rusbridger,

"The mass collaboration of strangers had achieved something it would have taken huge amounts of time and money to achieve through conventional journalism or law."

When a demonstrator was killed in a 2009 G20 protest in London, The Guardian asked anyone in the crowd who had photos or video to share them. The newspaper received evidence that police had shoved the man to his death. When thousands of redacted documents about the expense accounts of members of Parliament were made public, The Guardian put them all online and asked readers to help find the good bits. Rusbridger is making openness the hallmark of his news organization. "You cannot control distribution or create scarcity without becoming isolated from this new networked world," he says. "If ever there was a route to building audience, trust, and relevance, it is by embracing all the capabilities of this new world, not walling yourself away from them."[15]

In this regard, among others, Rusbridger puts himself in opposition to press baron Rupert Murdoch, head of News Corporation and owner of The Times, The Sun, and the now closed News of the World in London; the New York Post and The Wall Street Journal in New York; and newspapers in Australia.[16] Rusbridger sees the benefit in being open—and free—on the web. The internet expanded his reach worldwide. In the 1950s, 650 people in other countries read his paper. Today on the web, a total of 37 million people read The Guardian every day, two-thirds of them outside the United Kingdom. The Guardian, which is owned by a trust, is "profit seeking" but far from profitable. Murdoch, facing the same business pressures, is headed in another direction, building pay walls around his papers. "People reading news for free on the web, that's got to change," he said in 2009, and he soon willed it so.[17] He became the industry's chief advocate of the closed net.

It wasn't always so. In 2005, Murdoch addressed U.S. newspaper editors, celebrating the opportunity the internet gave them to "expand our reach" and to leave behind the "highly centralized world where news and information were tightly controlled by a few editors, who deemed to tell us what we could and should know." He challenged editors: "We have to free our minds of our prejudices and predispositions, and start thinking like our newest consumers. . . . They want to be able to use the information in a larger community—to talk about, to debate, to question, and even to meet the people who think about the world in similar or different ways."[18] That sounds like a strategy for social news. Only three months later, News Corporation bought the social home-page company Myspace for $580 million. People inside News Corporation tell me that his decision to acquire the company was practically an impulse buy made in a matter of days. Murdoch was then hailed as a web sage. Michael Wolff, author of the Murdoch biography *The Man Who Owns the News,* reported that as late as 2009, Murdoch had not used the internet or so much as searched Google. He nonetheless recognized the opportunity—and disruption—brought by this new technology.

Murdoch had already lost money on the internet. In 1993, he bought Delphi Internet Services, the first service to bring consumers to the net, through plain text.[19] I spent a topsy-turvy month there as head of content and quit just as Netscape introduced the first commercial web browser in the fall of 1994. News Corporation went on to squander a fortune on a failed and forgotten internet venture with the failed and forgotten phone company MCI. Murdoch sold Delphi back to its founders three years after buying it. Later, the internet bedeviled him again as Myspace faltered through many strategies and reorganizations. His newspapers were losing money online. The banking crisis caused a severe advertising implosion. In response Murdoch changed his rhetoric and strategy. He clamped a pay wall down around his Times in 2010. He pulled Times content from Google News, which he and his aides had demonized as—in descriptions compiled by Arianna Huffington—"parasites," "content kleptomaniacs," "vampires," "tech tapeworms in the intestines of the Internets," and thieves who "steal our copyright."[20] Les Hinton, then head of News Corporation's Dow Jones, complained to newspaper publishers, "We are all allowing our journalism—billions of dollars worth of it every year—to leak onto the internet. . . . They talk about the wonders of the interconnected world, about the democratization of journalism. The news, they say, is viral now—that we should be grateful. Well, I think all of us need to beware of geeks bearing gifts."[21]

Google has tried to school media companies in the new economy. "The large profit margins newspapers enjoyed in the past were built on an artificial scarcity: Limited choice for advertisers as well as readers," Google said in a paper submitted to the U.S. Federal Trade Commission in 2010. "With the Internet, that scarcity has been taken away and replaced by abundance. . . . It is not a question of analog dollars versus digital dimes, but rather a realistic assessment of how to make money in a world of abundant competitors and consumer choice."[22] No one has found the secret of sustaining news. I am working on new business models at my university and haven't cracked the puzzle either. I believe the answer will lie in finding new efficiencies through public collaboration and networks (with the partners formerly known as readers), building more engaged relationships with the communities news organizations serve (Facebook gets about thirty times as many page views per user per month as local newspaper sites), exploiting the value of those relationships in new ways (commerce and events, for example), and scaling local sales to serve small merchants in new ways (even helping them master Google and Twitter). The solution will not come from trying to preserve old business models.

Having failed to conquer the new world Murdoch heralded, News Corporation's executives are attempting to return to the old and familiar, the reality they once controlled. They used to be paid for their newspapers, so they insist they should be paid online. I have no objection to paying for content. I pay for cable, music, videos, books, Howard Stern on satellite radio, The Wall Street Journal online, and The New York Times in print—because each is in some way unique and valuable. The question isn't whether things *should* be paid for but whether they *can* charge consumers when their competition is free—and whether being closed or open will net greater profit.

News Corporation has not been public about the results of its pay wall. There were reports that the wall had cut The Times' traffic by 90 percent. In the early weeks, a reported fifteen thousand customers signed up to pay £1 a day or £2 a week for The Times online, in addition to those paying for a new iPad app. Meanwhile in the U.S., Murdoch started The Daily, a once-a-day newspaper designed for the iPad behind its own pay wall. Let's just say that walls will work, that they will pay the bills at The Times and The Daily and free them of dependence on fickle advertising. Let's just say. Still, the issue for News Corporation is that it has surrendered the larger public space to competitors, including The Guardian, Twitter, Google, Facebook, and The Huffington Post. News Corporation gave up on growth and on reinventing news for the new age. It acknowledged that it would only shrink in advertising revenue, which is the inevitable result of putting up the wall and cutting its audience so much. It declared defeat in understanding the business of publicness.

At this same time, News Corporation's Wall Street Journal embarked on an alarmist series decrying the privacy perils of online ad-targeting technologies—browser cookies and such—by media competitors and marketers. Mind you, The Journal itself uses these technologies and, because it has a pay wall, gathers more personally identifiable private data—name, address, and credit-card number—than many others do.[23] I'm generally not a conspiracy theorist, as I think organizations are rarely well-organized enough to conspire. But it's hard not to see this series as the company's shot across the bow of competitors who still depend on advertising. The series has the air of *Reefer Madness,* attacking web cookies as a gateway drug to privacy violations. The Journal practically invited government regulators—usually editorial anathema to the paper—to step in and control online advertising.

News Corporation is hardly alone in failing to reinvent itself. Other media companies thought they could buy internet strategies—Time Warner with AOL, CBS with CNET, NBC with iVillage, the New York Times Company with About.com, Condé Nast with Wired, the Washington Post Company with Slate—without upending their fundamental business structures. The music, publishing, and telecommunications industries are also desperately holding on to old models. Other companies have attacked Google and Facebook instead of learning lessons from them. Yet more have threatened to extend copyright or hamper the doctrine of fair use to disadvantage newcomers.

The problem for them all is that the old architecture of their markets has been razed by the internet and the publicness it demands. Media used to be built around brands. To get content, we had to go to the brand and buy its publication or watch its show. That put the media owner at the center, in control. Then came search, which flipped the relationship: Now, rather than starting with the media company, the transaction begins with the reader, who asks a question. If your content is there with an open answer—with the score to a game or the latest reports from a disaster—great; if not, you might as well not exist. Then came a force even more powerful: us. Our links, via Twitter, Facebook, blogs, email, and other social tools, now challenge the power even of Google. Content isn't king. Distribution isn't king. Relationships are.

But I'm a fine one to talk. Here I have put my content behind the pay wall known as a book. In *What Would Google Do?* I confessed my hypocrisy for not releasing it as a free, searchable, linkable online-only thing (or whatever we'll call the successor to the book). I did share and discuss many of the ideas in the book on my blog as I wrote it and after it was published. Now I must confess my sins with this book, too, and with the same excuse: Simon & Schuster paid me and gave me distribution in bookstores, not to mention considerable and valuable help from my editor. The system still works just well enough, and I took advantage of it. But I keep watching bookstores closing and monitoring the sales of digital books as they begin to eclipse those of print books in some categories. I'm concocting a plan for my next project, which would be created in public at events and online. The book, if there is one, would be a by-product and perhaps a marketing tool for more events.

That is the model Seth Godin is pursuing. The author of a string of best-selling business books announced on his blog in August 2010 that *Linchpin?* would be his last traditionally published book.[24] Godin can do that because he already has a strong brand. "I *know* who my readers are," he blogged. "Adding layers or faux scarcity doesn't help me or you." Godin has other revenue streams. He is in demand as a speaker, and no longer has to wait for agents to book him. He put together his own road trip of seminars, selling hundreds of tickets in each city for up to $895 for a full day, $200 for less time. When Meetup, a company that enables folks to organize in-person get-togethers, announced a new feature helping brands set up simultaneous Meetups around the world, Godin was the first to use it, and his fans scheduled more than a thousand events. Godin set up The Domino Project to produce and sell books directly through Amazon.com rather than a publisher—bypassing bookstores as well. The books are by-products of Brand Godin.[25]

Godin is to blame for my writing books. He sat me down one day and said I was a fool if I didn't write one—and I would further be a fool if I thought the book was the goal. No, he said, the book would build my public reputation, which would lead to other business. It has. I'm thinking I should follow Godin's example again and set up events of my own: seminars to brainstorm ways to gain value from publicness. Anyone game? (Email me.[26])

Retail: The social store

If you have a question for the electronics store Best Buy—"How do I hook up my this to my that?" "When are you selling the next new phone?" "Can you fix my broken printer?"—you can call and go through some level of phone-tree fun while enjoying the hold music. Or you can drive to the store and look for the right person to ask. Or you can go to Twitter and pose the question to @twelpforce.[27] That awkwardly named account—a contraction of "Twitter help force"—is monitored by three thousand "blue shirts" from the company's army of retail salespeople. When you ask a question, you can be assured that one of them across the country will have the answer in no time. I went to @twelpforce and asked about the best questions they've fielded. Derek Meister (@agent3012) responded (within minutes): "I always find the Twelpforce questions with very specific requirements interesting, like this one: 'What's the best TV for watching the *Blade* trilogy?'" Meister replied to that query with admirable specificity in a series of tweets: "Two things to consider about watching the *Blade* movies: 1) There's a lot of action, so you'd want an HDTV to be able to keep up with Wesley Snipes. That means a decently high refresh rate. 2) There's a lot of vampires, which means a lot of night scenes, and stark, dark colors. You want the shadows to be dark and the blood to be very, very red. So crisp colors, high contrast will help. Taking all that into consideration, you might want to look at plasma HDTVs. And not because of the obvious pun about vampires=blood=plasma. Samsung and Panasonic have some great options for you there." Another customer asked, "What kind of batteries do u recommend 4 a vibrator?" The intrepid @agent3012's answer: "Ummmm. Rechargeable ones."

Best Buy says @twelpforce drove down customer complaints by 20 percent in about a year. If those conversations had occurred one-on-one in the store, the rest of us would not have been able to share in the advice and the wit. But they occurred in public. At bbyfeed.com you can search all the answers the blue shirts have given customers. You can also go to Twitter and converse with Best Buy's CEO, Brian Dunn (@bbyceo), or its chief marketing officer, Barry Judge (@bestbuycmo). I met Judge on Twitter. He was nice enough to tweet about my last book, and I responded. That led to a few trips to Best Buy's Minneapolis headquarters, where I got to speak with executives and staff and witness the company's efforts to adapt to the opportunities and necessities of the public economy.

"It's a very different way of working than when I was younger," Judge tells me. "It was very warfare-like then. You had your secrets, and silence was good." But now Best Buy operates in ever more public ways. "At first it was a little scary because you thought you'd do something wrong," he says. But he acknowledges that the company is not exactly guarding "the secret potion for long life," so there's not much harm that can come from being open.

Besides, Judge says, "Every day we entrust our employees to interact with people across the world. Every day, they're out saying things. Essentially, that's all we're doing on the web." Trust them in the store, trust them online. The employees become the conduit from customer to company. Best Buy created an app that lets executives ask the blue shirts what they are hearing from customers. Blue shirts can also ask one another questions in all the stores. The employee, in this scenario, becomes "the human search engine." The company has experimented in Las Vegas stores with handing out preprogrammed tablets to blue shirts. Anywhere in the store, they are connected to more information and to experts from anywhere in the company. Judge suggested that customers could eventually use apps on their phones or be handed loaner tablets so, as they browse, they're connected with experts and each other—a glimmer of Josh Harris' vision of customers' social interaction around products.

On another front, Best Buy opened up its data. Outsiders can build their own applications and stores atop Best Buy as a platform. A bank created a store so its customers can use their points to buy merchandise via Best Buy. Camelbuy.com alerts customers to price drops at Best Buy. Milo.com searches three dozen chains' inventory—including Best Buy's competitors Target, Radio Shack, Sears, and Office Depot—to find the lowest prices for products. Best Buy is cooperating with the move to transparent pricing. It might as well. With my phone, I can already scan a bar code in the store and search online for more information and the lowest prices.

Transparent pricing is coming to most any retail category—even marijuana. PriceOfWeed.com compiles grass prices by geography, serving the average price per ounce for high-, medium-, and low-quality product. Because there is no grass superstore with a database API to call on, Price of Weed crowdsources the pricing information, asking customers to share what they paid—anonymously, of course. The site reveals how large the sample size is for each data point: the more information it gathers, the more accurate its averages will be. To prevent gaming, Price of Weed uses an algorithm that throws out outrageously high or low prices and sticks with the bell curve.[28]

Every company is holding on to data customers could use. In *What Would Google Do?*, I imagined the Googley restaurant, which would add data to its menu to tell us which dishes are the most popular and which wines are ordered with which entrées. I then tried to imagine the impossible: a Googley airline. I thought it would be based on openness and social relationships that draw out the wisdom of the passengers. Alas, no one has listened. Airline travel still sucks. But I have seen a few instances of opening up. Before its merger with United, Continental Airlines began to allow passengers to check on not only the status of their flights, but also the status of the incoming planes. That lets passengers make more reasonable judgments about whether the airline is ~~lying~~ being unduly optimistic. Continental also shows the waiting list of passengers standing by for seats and upgrades. Thus passengers know more about their fate without having to bug gate agents to share their secrets. There's no reason why that information should be hidden from us.

At the Burda DLD Conference in Munich in 2011, I emceed a workshop with Lufthansa and a roomful of social-media practitioners and media executives, speculating about the ultimate social airline. The frequent fliers in the room wanted the social airline to automatically give them the seats they like best. They wanted to pick their own seatmates. Offer us the food and drink we prefer, they said. Sync our seats with our iTunes playlists. The airline executives were surprised that these frequent travelers would hand over so much personal information and preferences, which would be golden data for the company. The travelers said they'd open up if they got benefits for doing so.

Could a new relationship with passengers turn an airline into a demand-driven rather than supply-driven business, finding out where people want to fly and flying them there? A bunch of KLM passengers ganged together in Twitter to ask the airline for a nonstop flight to a music festival in Miami. By organizing in public, they managed to fill 351 seats in only five hours and got their plane.[29] Changing airline schedules is highly complex, and airports are not set up for flexible routes. But Lufthansa already has a small private air service. That business could scale much larger if it could anticipate and fulfill demand. At the workshop, the owner of a social service for rich folks said he could fill a social flight every week.

How far does this openness go? Would Best Buy reveal, say, relative sales numbers of different television brands? On the one hand, the blue shirts do that already when they tell customers that one model is more popular than another. On the other hand, Judge says, it would be uncomfortable for the suppliers to have information about their sales revealed. Best Buy is the middleman in that transaction. For now. Transparency will disrupt that relationship. What can a retailer do when it can no longer arbitrage ignorance in the imperfect, opaque marketplace? Perhaps, Judge says, "business moves away from being so focused on hardware and more to content and services." Best Buy's Geek Squad will install and repair your equipment. The company is offering a buyback program: Pay a fee up front, and Best Buy will buy back your gadget for 20 to 50 percent of the purchase price. And content? Best Buy is already in the media business. Its newspaper advertising circular is media: manufacturers pay to be there. Manufacturers also pay for premium placement on store floors. In that sense, the store itself is an advertising medium. Best Buy gets 1.5 billion visits to its stores a year—more, it says, than to Times Square. If advertising surrounds Times Square, why shouldn't it surround Best Buy stores? Best Buy now has a staff selling presence in stores as advertising. It is trying to move past the business of selling stuff in boxes.

The bigger question Best Buy and other retailers have to ask is whether they should represent manufacturers to customers or customers to manufacturers. I'd like to see Best Buy bring out the wisdom of its crowd, involving customers in design projects as Local Motors does. Judge says the company is soliciting input for some of the house brands it sells. I think it can shoot higher: Best Buy could be the people's agent, taking their ideas and needs to manufacturers. If Best Buy's customers want a TV with ten HDMI ports; if they want the data-only cell-phone plan (because who talks on the things anymore?); if they want a laptop bag with a Kindle pouch, Best Buy can bring together these ideas and show they have critical mass. What manufacturer would ignore that market? Best Buy could engage its customers in a process of public design, ferreting out real demand and turning its relationship with the public on its head.

Going to extremes

The most radical vision for public business may be open-book management. John Case of *Inc.* magazine spotted the movement and coined the term in 1993[30] and later wrote a book heralding it as a business revolution.[31] The idea is that companies should share financial and performance data with employees so everyone can row together to the same goals.[32] It makes sense, but only if the employees see personal benefit—profit sharing, rising equity value, or, in the case of a start-up, success and survival—and if they understand the business dynamics.

Brian Golden, a professor at Toronto's Rotman School of Management, studied a lab-services company that opened its books—beyond just the bottom line—to get staff on board with workflow changes management was making in anticipation of a cyclical downturn. "The employees were scientists who didn't understand how to read the books," Golden is quoted as saying in Profit magazine. Instead of recognizing the downturn ahead, the staff saw no immediate problems. They demanded more pay and rejected management's moves. "With the staff/employer relationship poisoned," Profit reports, "the firm failed to prepare for bad times and its sales and staff count soon plummeted."[33] Oh, well.

Could a company open up its process of setting strategy? Surely that is best done behind conference-room doors in secret PowerPoint presentations. Or maybe not. The Wikimedia Foundation, which runs Wikipedia, undertook a year-long strategy project that engaged a thousand Wikipedians in producing twenty-six thousand pages of documents in a few dozen languages proposing and hashing over scores of ideas.[34] Sue Gardner, executive director of the foundation, told me they started the project aware of "a high likelihood of failure."[35] It was possible that no one would come to the party. Or the process could have been taken over by fringe interests and nutty ideas. The foundation hired a consultant to give the collaborators "a bedrock of information," and it hired a facilitator to guide the people and the process. In the end, the participants set priorities to expand Wikipedia in developing markets and to recruit more diverse authors and viewpoints in its developed markets.

OK, that may work for Wikipedia, which is already collaborative and does not face the pressures of profit a corporation does. Can the same process work for a company? There's only one way to find out. Wouldn't you like to hear where your customers want you to go next in a way that's less random and more engaging and empowering than holding focus groups? John Paton invited community members into one of his newsrooms and asked where he should go next. He doesn't have to do what they say. He's still responsible for the company's strategy and success. But now he has another way to listen. And it doesn't even require technology, only an open door and open mind.

Open-source endeavors also require leadership. When NYU's Jay Rosen researched open source, hoping to apply the ideas and experience to the news business, David Weinberger advised him that networks of people need wranglers. "The volunteers need their champion (and visionary)," Rosen says. Rosen went to meet Asa Dotzler, the people-wrangler at Mozilla, which produces the open-source Firefox browser. Dotzler told him that "anyone can play" is a key principle of openness. But quality also matters and needs its rewards.[36] Tasks are matched with the right person. Results are matched with the right incentives. There's still an economy in open source, but the currency is social—recognition, credit, status—rather than monetary.

Public broadcasting in the United States raises money from its audience (such volunteer contributions aren't as common in Europe and other countries because the public supports media through license fees [read: taxes]). Wikipedia and Mozilla also get capital from their users. David Cohn, an inspiring young journalist and entrepreneur, brought the model to local news with Spot.us, which allows individuals to pledge money to independent journalists to report specific stories about, for example, the slaughter of sharks in California and homeless people living in storage units.

Can this kind of public investment support for-profit endeavors? Kickstarter provides a platform that allows people to pledge money to would-be manufacturers, writers, filmmakers, programmers, designers, photographers, and performers. The people who give the money don't get equity or tax deductions. So why do a few users pay for others to benefit? They want to see the product made. And donors get special privileges. I gave $50 to support Scott Thomas, design director for the Barack Obama campaign, in his plan to produce the book *Designing Obama* without a publisher. He needed $65,000 to cover the cost of printing the 360-page coffee-table volume.[37] For my $50, I got a copy of the book, at a discount. Put another way, I preordered the book (923 people did likewise). If I'd given $10, I'd have had access to the digital version (117 did that). For $100, I'd have had my name printed in the book and gotten an embossed silver sleeve (that attracted 154 backers), and for $150 I'd have gotten a gold sleeve instead (113 did that). Altogether, Thomas raised $84,613 from 1,312 backers whose names are listed at Kickstarter (with links to their profiles, showing the other projects they back). Then he kept selling the book for $79.99. Every step along the way and all the stats I just listed are public, which makes the process a bit of a game as donors cheer for their guy to win and encourage friends to join the effort. Once he raised the money, Thomas kept his community of backers informed of his progress in designing, producing, printing, and distributing the book, even inviting people near him in Chicago to come pick up their copies at his studio.[38]

It's not unusual on Kickstarter for projects to receive more than they ask for. An ingenious tripod mount for the iPhone 4 called Glif asked for $10,000 but received $137,417 in orders.[39] I had wished such a thing existed. Now I could support it. For $20, I got the Glif when it was manufactured. For $50 I could have gotten one made sooner on a 3-D printer. The most successful Kickstarter fund-raising so far was by a designer who created two variations of a watchband for the iPod Nano. Scott Wilson asked for $15,000 to get started but received almost $1 million in orders.[40] So long as he planned his costs correctly, he has created a profitable pop-up enterprise using other people's production, distribution, and capital. The open Kickstarter model reduces risk. It demonstrates market demand. It allows entrepreneurs and artists to bypass old institutions and filters and get support directly from customers. With adjustment to investment law, I think this model should be a way to raise angel funding for new companies—truly public companies.

At the World Economic Forum in Davos in 2009, I helped organize a brainstorming session about rethinking industries around publicness and open collaboration. We gave the executives and entrepreneurs in the room a list of possible targets to tackle. They chose banking. In the midst of the financial crisis, they said, banking desperately needed disruption and repair. Their solution was the Open Bank. It would practice radical transparency: full disclosure of performance and compensation; collective risk assessment, allowing customers to help decide how to invest capital; open data to help customers compare their performance against others'. Bankers would not be allowed to sell a complex financial product unless they could pass a test proving they understood and could explain it. Customers could not buy the product unless they, too, showed they knew what it was. The company's tagline: "The Only Bank You Can Trust."

Publicness is also transforming other sectors, including education. Stanford, MIT,[41] and other universities have opened up their curricula, putting entire classes online for free. Why should thousands of teachers every year write the same lecture about analytic geometry when one of the best can be heard for free from MIT on iTunes? Public courses allow students to pick and choose the best teachers from anywhere. Those teachers get more recognition. Local teachers can concentrate on helping students understand concepts. Education becomes more efficient and scalable—which it desperately needs to be. See the amazing Khan Academy, which has hundreds of open lessons in math and science and has served them to students more than 43 million times in a new way to learn online. That is the beginning of a seismic disruption of education.

Think, too, about the radically public foundation. Rather than making decisions in private, why not open up the search for grantees, their selection, and the monitoring of their success? The John S. and James L. Knight Foundation holds an annual challenge to attract hundreds of proposals to update the news industry. Many of the proposals are put online, openly, to solicit ideas and suggestions. One issue with this approach is that the foundation attracts more proposals and has to say no more often. On the other hand, it can attract ideas it wouldn't have found via the old, closed system. I direct the Tow-Knight Center for Entrepreneurial Journalism at CUNY (funded in part by Knight), and I hope to use this principle to manage support of new ventures. I'm inspired both by Knight and by new, for-profit investment funds—Y Combinator, for example—which hold open competitions among start-ups to award the best of them funding, incubation, and education. In foundations and venture investing, the process of giving out money used to be secret. Now both are opening up.

The best thing about writing *What Would Google Do?* was hearing afterward from people who applied its principles to their enterprises, taking the ideas further than I could have imagined. I read blog posts by pastors speculating about the Googley church.[42] They liked the idea of churches as betas. I spoke with managers of senior citizens' communities who wondered how they could more generously share their knowledge with families before parents moved in. I spoke with hundreds of truck-stop owners in Las Vegas and imagined how they could create social and information networks among their drivers. Here I've given just a few examples of how the principles of publicness can be applied to companies. You will imagine more and more radical ideas, and you can tell me what works and what doesn't. I urge you to share your thoughts openly at www.buzzmachine.com/publicparts so the brainstorming can continue.

By the People

Killing Secrecy

Government should be open by default, secret by necessity. Instead, too much of government is secret by default, public by force. I've joked—well, half-joked—that the U.S. Freedom of Information Act should be repealed and turned inside out so we no longer have to ask government to open up our information. Government should instead have to ask our permission to keep it from us. There are still good reasons for secrecy: security, defense, criminal investigations, privacy, and certain matters of diplomacy. The rest is our information. In addition to appointing privacy czars to protect citizens' information and security czars to protect governments', we should have publicness czars who represent the interests of the people in openness.

If the government doesn't have its own agents of openness, the public will. WikiLeaks is such an agent. It has changed governments' expectation of secrecy. Since 2006, WikiLeaks has released documents about Icelandic and Swiss banks, Scientology, corruption and human-rights violations in Kenya, chemical dumping in Africa, as well as Congressional Research Service reports that hadn't been available to citizens.[1] WikiLeaks' impact and influence leaped higher in 2010 when it released leaked video of U.S. helicopter gunships in Iraq shooting people who appear to be civilians, including Reuters news agency employees. WikiLeaks founder Julian Assange called it "collateral murder."[2] Next came thousands of documents about the Afghanistan and Iraq wars, then a trove of U.S. State Department cables that had been available to thousands of government employees and were leaked by one. A selection of the documents—though not all—were opened up to citizens and the world with the help of media organizations, including The Guardian, The New York Times, and Der Spiegel. Those journalistic organizations redacted sensitive information, added context, and promoted the leaks, which is why Assange partnered with them—to get more impact. Commentators debated how surprising the information in the leaks was. The cables revealed small embarrassments—a diplomat criticizing Germany's Angela Merkel or Russia's Vladimir Putin. They also revealed, some commentators said, the quality of the diplomatic corps' work. Cables about Tunisia only confirmed citizens' already well-founded suspicions about corruption there, it's true. But those cables became a catalyst of protests that led to the government's downfall, which in turn led to the storm of protest that soon overtook the Middle East.

On the whole, the leaks demonstrated the banality of secrecy. They showed that government keeps too much secret. The leaks also taught government what it will be like to operate under a presumption of publicness. NYU's Clay Shirky formulated a law to calculate the vulnerability of secrets. In an organization, he says, any given secret is as secure as the most adverse cost-benefit calculation by the least committed user. A hospital clerk, for example, sees a document revealing a star's pregnancy. If the clerk sells this information to a tabloid, he'll lose his job. But he'll make more money in a day from the leak than he makes in a year from his job. And he hates his job. Cost-benefit good. Commitment low. He leaks. Security expert Bruce Schneier states Shirky's law more simply: "Secrets are only as secure as the least-trusted person who knows them."[3]

That dynamic is hardly new. Daniel Ellsberg leaked the Pentagon papers to The New York Times long before the internet. He had to spend months photocopying the documents and sneaking them out of government offices. What's different today is that computers and networks make more information available to more people so it can be found, copied, and shared more quickly. The internet provides an architecture of anonymity, which WikiLeaks founder Julian Assange exploits. He also exploits the expanding ethic and culture of openness I've chronicled in this book. Together these changes challenge power, shifting influence from presidents and government ministers to the likes of Assange. He sees openness as a way to redistribute control. "Transparency should be proportional to the power that one has," he says.[4] There lies a second law, the one that drives Assange: Those who held secrets once held power. Now those who create transparency gain power.

In The New Republic, Noam Scheiber predicted that there will be two basic responses from governments and businesses to WikiLeaks' disruption: First, they can shrink into smaller organizations where fewer people know secrets, so the opportunities for leaks are reduced. Second, they can tighten security. That was the reflex of the U.S. government, which went on the attack against WikiLeaks. Vice President Joe Biden called Assange a "high-tech terrorist." Bradley Manning, the young soldier accused of feeding WikiLeaks the helicopter video, war documents, and cables, was held in solitary confinement. Senator Joe Lieberman pressured Amazon.com into throwing WikiLeaks off its servers. Officials reportedly pressured PayPal and banks to stop funneling donations to WikiLeaks.[5] Authorities can try to play catch-up with leaks, but that will prove frustrating—especially as long as they can be accused of keeping too many secrets. They have lost their credibility to classify information. Classifying more is not the answer. The opposite is true. The more secrets you keep, the less you are trusted to decide what is secret and the more leaks there can be. And there is a third law: The only sure defense against leaks is transparency.

The truth will out. In his book WikiLeaks and the Age of Transparency—an excellent survey of the new political landscape around openness—Micah Sifry catalogues projects that use databases, crowdsourcing, networking, social interaction, mapping, and other internet technology to make government open in spite of itself. The Sunlight Foundation, led by Ellen Miller, underwrites OpenCongress, which maintains wiki biographies that track campaign contributions, legislation, and votes for each member of the House and Senate. It backs MapLight.org, a research tool to find correlations between contributions and legislative votes. Similar tools, at OpenSecrets.org and FollowtheMoney.org, also follow campaign money. Sifry tells the story of linguistics graduate student Joshua Tauberer, who built GovTrack.us to scrape open data off government servers and convert it to formats developers can use to analyze it. The U.K. is blessed with many such data activists. mySociety, run by Tom Steinberg, champions open data and builds applications to use it. TheyWorkForYou .com lets citizens track the activity of members of Parliament. WhatDo TheyKnow helps people file Freedom of Information requests.

Technology in itself isn't sufficient. Reform needs the effort of citizens. One of my favorite transparency projects is Porkbusters, a libertarian/ conservative effort to uncover government waste. It joined forces with liberal blogger Joshua Marshall at Talking Points Memo when a secret Senate hold was put on legislation aimed at building a database of federal contracts and grants. This tripartisan group had its readers call senators' offices one-by-one to ask the simple question: Did you put the hold on the bill? The process of elimination turned up the heat on the culprits, the late Senators Ted Stevens and Robert Byrd, who confessed and let the legislation go through. The bill became law.[6]

We would be unwise to make transparency just a game of government gotcha, of demanding information only so we can catch the bastards red-handed. Oh, there is no shortage of red-handed bastards to catch. But Harvard Law School professor Lawrence Lessig worries that if we use transparency exclusively to expose misdeeds, without the benefit of context and understanding, we "will inspire not reform, but disgust."[7] He's not against transparency. He wants to solve the problems it exposes in other ways (by reforming campaign finances before contributions can lead to corruption, as one example). I share his trepidation. If all we try to do with government information is fight government, government will fight transparency.

Openness does not merely expose misdeeds, it exposes value, making open data an economic imperative. Lessig says that making weather data public "produces more than $800 million in economic value."[8] Sharing government-financed global positioning data led to the creation of now-indispensable navigation systems and mapping on our smart phones, which in turn powers such services as Foursquare. Agricultural data save and make money on farms. Data can lead to the creation of new companies. Writing on the blog TechCrunch, executive-turned-academic Vivek Wadhwa tells the story of entrepreneurs who got government to release data on retirement plans so consumers could compare and shop. "There are literally thousands of new opportunities to improve government and to improve society—and to make a fortune while doing it," Wadhwa says.[9] "There is a new compact on the horizon," declares publisher and tech visionary Tim O'Reilly in the book *Open Government*. "Information produced by and on behalf of citizens is the lifeblood of the economy and the nation; government has a responsibility to treat that information as a national asset."[10] So cough it up.

Carl Malamud, who runs PublicResource.org, is a hero of the open-government movement.[11] In the 1990s, the U.S. Securities and Exchange Commission would not make its EDGAR database of public companies' filings available for free. If you wanted it, you had to trek to a government reading room or subscribe to an expensive private database such as Nexis. In 1993, Malamud received a $600,000 grant from the National Science Foundation to buy the data and put it on a server donated by Eric Schmidt, then chief technology officer of Sun Microsystems. Malamud democratized the data, making it available to journalists, investors, and entrepreneurs as well as corporations and traders with deep pockets. Rather than seeking more funding to keep his service afloat, Malamud offered his technology—along with training—to the SEC for free. He shamed it into taking the service over.[12] In the meantime, Malamud has posted more than four thousand videos owned by the government to YouTube and FedFlix. In 2010, Google's philanthropic arm, Google .org, granted Malamud $2 million (just in time, as PublicResource.org was down to its last $2,000) to help local governments put their building codes online and to fund other projects.[13] He is also lobbying for a change in Freedom of Information Act procedures, so that when one citizens' request is granted, the results are posted online for all to see.

A group of Princeton computer scientists wrote a 2009 paper arguing that rather than making web sites, government agencies should release data in standard formats so that anyone—including the agency itself—can analyze the information and make web sites and applications from it, creating competition to improve the presentation of information.[14] Outsiders, they argue, will do a better job of making data useful as they are less hampered by the government's own regulations. They can perform some functions better than government and save government the effort and money. But first the data need to be available to developers in a usable form. New York City has made three hundred fifty data sets public at nyc.gov/data, including building complaints, a calendar of festivals, data about 9/11 survivors' health, and restaurant inspections. The U.S. government has released almost three thousand data sets at Data.gov, including campaign finance records, food and drug recalls, and a government purchases database.[15] The U.K. government has released more than four thousand data sets at data.gov.uk.[16] In the U.K., data liberation activists had to fight to get the government to free up mapping data. The government had viewed the data as a source of revenue, but the activists said it was the property of the people, who'd already paid for it with their taxes. Eventually, the U.K. adopted an Open Government License to make its intellectual property—its data and its software—open and free for reuse.[17] Its goals: to promote innovation, to make government transparent, and to facilitate civic engagement. But the battle to liberate data continues. Hard as it is to believe, the taxpayer-supported Smithsonian Institution and some U.S. states still try to copyright their data. Oregon, for example, claimed ownership of the state's laws until it relented in 2008.[18]

Once data can be scooped up and analyzed, government and citizens can move past transparency to the next step: collaboration. Wise governments ask programmers to turn their data into applications. New York City has held hackathons—one with $20,000 in prizes—to make the city "more transparent, accessible, and accountable" through writing applications around restaurant inspection reports, traffic updates, traffic cams, directions, consumer complaints, library catalogues, maps to playgrounds, home prices, lists of most-wanted criminals, and a city tree census (142,747 trees grow in Brooklyn[19]).

Beyond Openness

Beth Noveck wants collaborative governance. She doesn't much like the phrase "open government." But it's her own fault that those words became part of her job title when she headed President Obama's Open Government Initiative until 2011. Noveck wrote a book called *Wiki Government* about collaborative democracy. She worried that if she used the term "collaborative" to name the White House endeavor, it would be perceived as an effort to promote her book. "It's like the time I voted against myself for fifth-grade student council, and I lost by one vote," she says. So she settled instead on "open." But the word doesn't go far enough. It doesn't say what comes next after opening—namely, doing.

When I sat down with Noveck to talk about her work, she set the stage by repudiating our friend Habermas: "We need to move away from the model of a single public sphere in which everyone participates equally in the same way. So the traditional, deliberative conception that there is some great salon in which everyone is participating is the misconception." Echoing the Making Publics project, she says the internet allows us each to "join our own public that speaks to our own passions and our own sense of meaning. . . . It's allowing people to *do things*."

Noveck wants a structure of collaboration that embraces our different abilities, interests, and agendas. She likens it to community organizing. "One person gets the coffee and another person distributes the leaflets and a third person makes the phone calls. It just speaks to people's different skills and talents." She also wants to "blow open the concept of what we mean by expert. . . . We have defined our concept of expert to limit it to a small number of people. What the internet does is treat everyone as an expert." Well, not *everyone*. But *anyone* can prove to be an expert, as Wikipedia demonstrates. Noveck wants to bring that spirit to government with "institutions that take advantage of distributed expertise, distributed passion."

In her book, Noveck recounts a project she worked on before joining the White House, to bring new expertise to the slow (not to mention anachronistic and dangerous) U.S. patent approval process. The Patent Office employs 5,500 examiners, most of whom are not experts in what they are assessing. They face a gigantic backlog and a lag time of at least three years in a technology environment that considers three years a generation. "All this got me to thinking, what if the patent examiner worked with the broader community?" Noveck asks. "What if the public augmented the official's research with its own know-how? What if the scientific and technical expertise of the graduate student, industry researcher, university professor, and hobbyist could be linked to the legal expertise of the patent examiner to produce a better decision? What if, instead of traditional peer review, a process of open review were instituted, wherein participants self-select on the basis of their expertise and enthusiasm?"

Thus was born the Peer to Patent pilot project. It published selected patent applications openly, attracting experts to help research them. And it's working. The Patent Office is adopting the project, and the idea is spreading to other countries. I'd have thought that patent examiners would resist interference, but Noveck says they enjoyed talking with experts in the field. Their jobs used to be lonely. This project made their jobs social, not to mention more efficient. In a speech to the New America Foundation, Eric Schmidt used Peer to Patent as a model: "Why is that not true of every branch of government? It makes perfect sense: Use all those people who care so passionately and who have a lot of free time to help you."

The National Archives did just that to rethink and redesign its Federal Register. Called the daily newspaper of the federal government, the Register publishes notices and proposed and enacted rules of federal agencies as well as executive orders. It set a landmark in government transparency when Franklin Roosevelt ordered it compiled and published in 1935. Before then, each agency archived its own rules. Finding even the published actions and notices of government was difficult—and much was not published.[20] In the U.K., until the eighteenth century, Parliament did not allow the press to report its proceedings.[21] In the United States, until 1800, Congress did not admit reporters to its sessions.[22] As late as the 1950s, more than a third of congressional committee meetings were held in secret.[23] The Register made much of government's news available. It did not, however, make it broadly and easily accessible. The Register is a festival of fine print, not easy to digest. In 2009, the Sunlight Foundation held a competition to improve it. Three computer-savvy citizens in California thought they could make it look better. "They don't even know what the Federal Register is," Noveck says. "What it is to them is the biggest data set in Data.gov." The developers—Andrew Carpenter, Bob Burbach, and Dave Augustine—used Register data and open-source software to create GovPulse.us. On the site, citizens can explore documents by agency or topic, receive alerts when comment periods are closing, and find documents that mention their towns.[24] The site does the impossible: It makes government data cool. The developers won second prize, but the National Archives and Records Administration was so impressed that it hired them to use the same data to build Federal Register 2.0 at FederalRegister.gov. "In three months they create the new newspaper of our democracy," Noveck says. "USA.gov meets USA Today."

Noveck would like to bring that kind of effort to other government agencies. "Social Security touches the lives of 60 million Americans a month," she tells me. "It's a quarter of the federal budget. And yet it has a web site that is in English and is entirely inaccessible to low-English-proficiency people." She suggests crowdsourcing translation. Facebook translated its own platform into sixty-four languages (with thirty-two more in process)—with its own users.[25] Translating the Social Security site could be done on Facebook with a group for every language taking on its dozen most important sections.

Collaboration is a two-way wire. Government is not the only source of data. Outsiders can also give data to government to help it work better. Ushahidi is a platform built for citizens to gather their own information with their mobile phones. It has been used to track response and recovery efforts after the earthquake in Haiti;[26] snowplowing problems in Washington, D.C. (helping city officials target problem areas);[27] and voting incidents in Ethiopia, Namibia, and India.[28] People know what is happening in their communities. Ushahidi helps them gather, share, and analyze their own information. Then government can act.

These examples are all built—could only be built—on open data. Citizens use government as a platform, as it should be. They bring unique skills, expertise, and passions to the tasks. They solve problems that matter to them—not necessarily the issues government or the press identify. Open government isn't as simple as releasing data or holding hearings. In these examples, there is often a progression in openness from simpler to harder tasks, from

1. transparency—opening up information as a platform for action—to

2. identifying problems—taking advantage of new ways to listen to the public—to

3. convening parties to address those problems—something the president or the owner of a press can still do well—to

4. identifying solutions—which takes expertise—to

5. executing—which often still requires the authority and resources of government.

But not always. Government need not be at the center of identifying and solving society's every problem. It cannot afford to be. Its work is expensive, slow, and hobbled by legal and political limitations, not to mention pressure from special interests and bouts of corruption. And it's hard to innovate inside government. Government is hardly a beta culture like Silicon Valley. "Failure is not an option," Noveck says. There is a real risk if government messes up in matters of security, health, and privacy. Noveck looks instead for ways to share both the work and the risk with outsiders. "We do some things at arm's length," she says. "We'll come up with a concept of how to do something. We'll work with a company, a foundation, a third party, to let the thousand flowers bloom, and then we can cherry-pick out of that. It preserves the integrity of the government player when we work with a partner." In other words, she lets outside organizations be government's laboratory.

That's one way to get around politics and improve government. But the bigger question is: Can we also use the tools of publicness to improve the political process? That's where even I, internet triumphalist and incurable optimist, become gloomy. As long as we still have mass media, we'll have mass advertising in campaigns and big money's support of them—and we'll still see their corrupting influence on political discourse. The institutions of power aren't motivated to change. They have the means to stop change. That would be a tragic irony: Our tools of publicness shift power in many sectors across society—except in our public sphere. Habermas was right to lament mass media's impact.

Would it be possible to elect web candidates who use the tools of publicness—the ability to listen, spread messages, organize followers, and create deeper, closer relationships with constituents—to disrupt the parties? I've seen no evidence of that yet. Yes, Barack Obama raised a fortune online, but where did that money go? To TV. In Europe's parliamentary governments, upstart parties can gather attention but little influence (the Pirate Party has two members of the European Parliament from Sweden, otherwise just city-council seats in the Czech Republic, Germany, and Switzerland).[29] When I asked about this dilemma on Twitter, someone said that The Movement in Iceland began with blogs and Facebook. But then on that small island one could organize a movement with a mimeograph machine and megaphone.

It's tempting to see the U.S. Tea Party as a movement made possible by the new tools of public organizing, but I think that would miss the mark. The Tea Party was built more through the power of talk radio and Fox News. It is still the child of mass media, not social media. Then again, Micah Sifry argues that the Tea Party is new in this sense: It is a network rather than a top-down political structure. Like the youth revolts in Egypt and Tunisia, the Tea Party seems to have no clear leadership structure. "Instead of having one central address or leader, it has lots of small groups and no real barrier to entry," Sifry says. The party has only ten full-time staffers but claims thousands of local organizations and millions of adherents. "Thus no one can control it or stop it by delegitimizing a single leader."[30] Perhaps that is a new, networked model for organizing movements to win elections. But it's too soon to tell.

Once elected, a movement's new representatives will still be stuck in the molasses that is Washington. Or, as in the cases of Tunisia and Egypt, we shall see whether the networks that toppled governments will be capable of organizing governments. "The shift of balance from the institutional society to the network society will topple dictators, bring down governments, occasionally create terror and mayhem, create economic risk and opportunity, and quickly eliminate some traditional civic and state institutions," writes U.K. Labour politician Anthony Painter. "Our success as a movement is determined by our ability to build enduring institutions of change out of networks of outrage."[31]

Perhaps we might try building shadow institutions. What if outsiders organized, for example, a parallel Federal Communications Commission that would attract its own experts, set its own priorities and policies, and analyze government data—once open—independently, apart from think tanks and lobbyists? Would it be possible to have an FCC of and by the net culture as a counterweight to the power of government? Can we create new public spheres that act to balance government, meeting not in Habermas' coffeehouses but online? If we want to try, we need to do more than merely comment and complain.

We need a new sense of what the public is, independent of government and media. Perhaps we see this new sense of a public blooming in Egypt and Tunisia. There and in other nations, citizens are using the tools of publicness and creating networks not to fill the odd pothole but to give birth to freedom and to governments of and by the people. As I witnessed these movements, I reflected on my own assumption that advances in digital democracy would more likely come first in the West, where our democracies and our digital societies are more advanced. I was wrong. Those advances are more likely to come where they are needed most. The Middle East may end up leading the way in rethinking what it means to be a public.

I have also witnessed a debate over the impact of the tools of publicness on civil society. Technological curmudgeons are waging that argument with so-called technological determinists. The alleged curmudgeons include *Net Delusion* author Evgeny Morozov[32] and New Yorker writer Malcolm Gladwell;[33] the alleged determinists, New York University's Clay Shirky,[34] The Daily Beast's Andrew Sullivan, and me. Our critics say we give too much credit to social tools in these revolutions. I know no one who argues that the tools cause or carry out revolution. But they have facilitated and perhaps hastened it. Amusingly, the exact same argument still roils among Gutenberg scholars about the impact of the press as a tool of social change five centuries ago. Elizabeth Eisenstein is labeled a technological determinist by her critics, including Adrian Johns, author of *The Nature of the Book* [35] and a member of the curmudgeonly camp Eisenstein calls catastrophists. Johns does not buy her contention that print was itself revolutionary and "created a fundamental division in human history."

It is too soon to judge the full impact of the next Gutenberg transformation, the one we are experiencing now. I believe the changes are only just beginning. It took seventy years for the printing press to serve the cause of Martin Luther's reformation. John Naughton, a columnist for the Observer in London, asks us to imagine we are pollsters in 1472, seventeen years after the first printed Bibles (we are only that far away from the invention of the web today). We are on a bridge in Mainz asking citizens how likely they think it will be that Gutenberg's invention will:

(a) Undermine the authority of the Catholic Church?

(b) Power the Reformation?

(c) Enable the rise of modern science?

(d) Create entirely new social classes and professions?

(e) Change our conceptions of "childhood" as a protected early period in a person's life?

"Printing did indeed have all these effects," Naughton says, "but there was no way that anyone in 1472, in Mainz (or anywhere else for that matter) could have known how profound its impact would be." [36]

The New World

Who Will Protect Publicness?

Who will protect today's Gutenberg press—our tool of publicness—the internet? Governments? Companies? Someone must. During Egypt's revolution, the doomed Mubarak regime killed the net. At the same time, China prevented internet users from searching services such as Twitter for the word "Egypt." That is the least of what China does to restrict internet freedom. It censors sites and search behind the all-too-whimsically named Great Firewall of China, maintained with the help of U.S. technology companies. It outlaws entire services, such as Facebook, that can be used by citizens to organize protests, which the government propaganda office then decrees cannot be reported.[1] It uses the net to find and arrest dissidents, once with the help of Yahoo. It uses the web, too, to spread disinformation. Governments are not the net's proper guardians.

In 2009, Chinese hackers attacked Google and other technology companies. The Chinese government has not acknowledged responsibility. But in a diplomatic cable released by WikiLeaks, a U.S. envoy's sources said a top member of the Chinese politburo orchestrated the assault. He was upset at what showed up under a Google search for his name.[2] Google publicly revealed the hacking. It also used the opportunity to reverse its policy on China. Google had been censoring searches, blocking some results about such topics as Falun Gong and Tiananmen Square. Now it refused to continue censoring search on behalf of the Chinese government. It threatened to pull out of the country. Some company executives had believed that offering a hampered Google to China was better—for the Chinese and for Google—than having none. Others believed it was wrong to do the dictators' bidding—and they at last prevailed. Google defended its principles.[3] At the time, no other corporate victim of the hacking stood by Google, and no government was willing to lend its vocal support. Eric Schmidt told a group of editors in 2010 that Google did not have laws or police or diplomats. It is not a country, he said. But in this case, it acted as a quasi-state, as the internet's ambassador to China. Is Google the net's protector?

At the Aspen Ideas Festival in 2009, Schmidt defended "the openness of the internet."[4] Yet a year later, Google entered into a devil's pact with telco Verizon over the net's openness. Proponents of net neutrality want to see all internet traffic treated equally. Internet service providers want the ability to discriminate against some kinds of traffic. They argue for the power to govern alleged bandwidth hogs who download more than others. They might also want to favor their own services over a competitor's. Google and Verizon made a recommendation to the Federal Communications Commission to split the baby with a dull knife, and the FCC followed much of their suggestion. It divided the internet in two: the wired internet (the past) would have neutrality protection, while the wireless internet (the future) would have less. It's the internet vs. the schminternet. If you use your iPad at home, connecting to the net via your cable company's wires, then walk outside and reach the net on the same device via your mobile carrier's signal, you'll be operating under different regulations. The movie you're watching could slow down just because the rules are different. Welcome to the schminternet. The debate over net neutrality is often framed in those terms: downloading movies at full speed. But there's more at stake here. Whether an internet service provider cuts off movie downloads or China cuts off searches or Egypt turns off the entire internet, the question is the same: Who decides which bits get through? Who defends the openness of the internet? Who then nurtures the opportunities we are just beginning to understand? Can companies be the net's protector? No.

"Changes in the information age will be as dramatic as those in the Middle Ages," James Dewar writes in a 1998 Rand Corporation paper. "The printing press has been implicated in the Reformation, the Renaissance, and the Scientific Revolution, all of which had profound impacts on their eras; similarly profound changes may already be underway in the information age."[5] Dewar argues that the information age will be dominated by unintended consequences. The wise course, he says, is not to forestall these changes with regulation and resistance but to hasten the path and the adjustment to them. This transformation will be unsettling, sometimes frightening, and monumentally disruptive to the hierarchy of society. Corporations, governments, and old institutions are not built to operate so much in the open. Oh, they could change—and some will— but the transition requires a wrenching and expensive shift. By the time they recognize the need for change and the inevitability of it, new entrants using open platforms will often be ahead of them. Amazon.com disrupted retail faster than retail could shift. craigslist sneaked up on newspapers and shifted $13 billion a year from the companies' coffers to consumers' pockets. The regimes in Tunisia, Egypt, and Libya were incapable of re-forming in incremental steps; their citizens decided to start over.

The tools revolutionaries and disruptors use to tear down the old order may not be sufficient to build a new one. It is also true that the tools are neutral—they can be used by bad actors as well as good. After the Egyptian revolution, three CDs filled with ID photos of Egyptian security police were found in security headquarters. The photos were put online at Flickr so citizens could crowdsource identifying them. Flickr took them down.[6] NPR's Andy Carvin asked why and was told that they violated the site's terms of service, which require users to post original photography, and that an unnamed user—a government?—had complained. Then Anonymous, a corps of hackers that often defends WikiLeaks, took the photos and put them online again.[7] The truth cannot be stopped. In that instance, the tools were used for good. But Evgeny Morozov also tells the story of security forces in Iran who took photos from demonstrations and put them online so loyalists could identify protestors via crowdsourcing and police could then arrest them. He speculated that facial-recognition software could be used to do the job instantly. These tools can be used for evil purposes as well.

Should we regulate and control the internet so it can be used only for good? Rand's Dewar says that early modern European nations that tried to control the printing press and suppress its allegedly dangerous aspects not only failed but also fell behind other neighboring nations. "It was more important to explore the upside of the technology than to protect against the downside. In the information age, this suggests to me that the internet should remain unregulated," he writes.[8]

I fear the unintended consequences that may come from regulation. Take, for example, European Union Justice Commissioner Viviane Reding's four pillars of data protection, which she proposed in 2011.[9] I have no argument with one of them: transparency. Companies that collect data should be open about when that is done and how information will be used. Another pillar sounds attractive: "the right to be forgotten." But how far does that go? If I post something about you on my blog or write about you in a news story—a quote I heard, the fact that I saw you somewhere, the fact that you did something in the open—can I be forced to erase—to forget—that? What then of my freedom of speech? Another pillar is rhetorically appealing: "privacy by default." But is that how we wish society to operate—closing in by reflex when we have so many new ways to open up? Flickr became a success, as I said earlier, because it was set to public by default. On a service designed for sharing—Facebook— what does complete privacy mean? Isn't completely closed communication just email? Reding's last pillar would require EU-level protection no matter where a service operates or where data are held. That sets a dangerous precedent. It could mean that we would all be ruled by the most stringent controls in place anywhere in the world—the high-water mark of control.[10] Can we bear China claiming the same right as the EU? We see a related problem today with so-called libel tourism in the U.K. Because its libel laws are unfriendly to defendants, targets of published criticism go there to file suit against writers and publishers. In a global internet, the EU's effort to become privacy's sanctum could affect us all.

On the one hand, I argue against regulation. On the other hand, I argue that the government should enforce net neutrality, and that is a form of regulation. Am I hypocritical? At South by Southwest in 2011, Senator Al Franken delivered a ringing endorsement of net neutrality. He argued that proponents of net neutrality are not trying to change the internet but to keep corporations from changing it, from making the net less free than it has been since its birth. "This is a First Amendment issue," he said. "The internet is small-d democratic. Everyone has the same say."

Secretary of State Hillary Clinton, too, delivered a rousing defense of internet freedom in two speeches in 2010 and 2011. "In the last year, we've seen a spike in threats to the free flow of information. China, Tunisia, and Uzbekistan have stepped up their censorship of the internet," she said in Washington just as the Tunisian revolt was brewing. "On their own, new technologies do not take sides in the struggle for freedom and progress, but the United States does. We stand for a single internet where all of humanity has equal access to knowledge and ideas. . . . The internet is a network that magnifies the power and potential of all others. And that's why we believe it's critical that its users are assured certain basic freedoms. Freedom of expression is first among them."[11] The following year, in 2011, she delivered another speech extolling transparency and attacking censorship. But in the same speech, she also condemned WikiLeaks for its release of cables from her agency. "Let's be clear," she said, "this disclosure is not just an attack on America—it's an attack on the international community." The leaks "tear at the fabric" of government, she argued.[12] Indeed, they soon tore at the fabric of Tunisia's corrupt government.

We cannot rely on governments—neither democracies nor tyrants—to safeguard the tool of their own disruption. Nor can we expect corporations—not Google, not cable, not telephone companies—to operate contrary to their own self-interest. "Big corporations? They're not inherently evil," Franken said at South by Southwest. "Corporations have a contractual duty, a legal obligation to their shareholders to make as much money as they can." They can make more money by putting some content in the fast lane, relegating the rest to sit in traffic, and favoring their own services—the networks Comcast bought with NBC Universal, for example—over competitors'.

I do not blame them for protecting their own interests. I blame us, the people of the net, for not protecting ours, for leaving the internet vulnerable and jeopardizing the disruption of the old order and the development of our new and open society. Where is our protest at the violation of the Chinese people's fundamental human rights to speak and assemble? When Chinese citizens, following the example of peaceful movements in the Middle East, used the net to organize events and simply stroll together at certain landmarks, government security forces came to stop them. Did we stand up for them? Where is our concern over the implications of government attacks on WikiLeaks as an agent of transparency? Would we have stood for such intimidation of The New York Times and Daniel Ellsberg? Where is our dismay at the diminishment of the public domain with the Germans' *Verpixelungsrecht*? Where are our principles?

As I contemplated those questions, I thought back to the Sullivan Principles, which helped put an end to apartheid in South Africa.[13] To be clear, I do not equate the repression and tyranny of apartheid with the censorship of YouTube videos or the throttling of music downloads. But there's a lesson in the Reverend Leon Sullivan's proposition that companies must operate responsibly. In 1977, he wrote a set of principles to use as a wedge to pressure companies doing business in South Africa to ensure equality for workers or get out. Business, companies, and capital left the country in a process that helped freedom emerge there.

We need principles to defend our internet and our publicness—fundamental beliefs that we, the users and citizens of the net, can point to when we see governments or companies violate them and threaten our freedoms. I don't want to establish a set of laws made by the United Nations or the governments that make it up. I want principles independent of those authorities. For the tools of publicness are what empower us to check the powerful. I quote the magnificently over-the-top Declaration of Independence for Cyberspace written by John Perry Barlow, a founder of the Electronic Frontier Foundation (and Grateful Dead lyricist), in 1996: "Governments of the Industrial World, you weary giants of flesh and steel, I come from Cyberspace, the new home of the Mind. On behalf of the future, I ask you of the past to leave us alone. You are not welcome among us. You have no sovereignty where we gather. . . . Governments derive their just powers from the consent of the governed. You have neither solicited nor received ours."[14]

Barlow taunts governments, saying they do not know "our culture, our ethics, or the unwritten codes that already provide our society more order than could be obtained by any of your impositions. . . . We are forming our own Social Contract."[15] He's right. Among users of goodwill, we see a force of collective effort to create order, propriety, and justice in the digital world. A new and good society is emerging. We don't yet know the shape of it. I have tried to use Gutenberg as a frame of reference. Others use economic terms. "Personal data is the new oil of the Internet and the new currency of the digital world," says Meglena Kuneva, the European Union's consumer affairs commissioner.[16] Marc Davis, a researcher at Microsoft, uses the construct of property. He argues that we do not own the data we create, the data about us, the servers on which our information resides, or the wires that bring it to us. He believes we need a set of property rights and trusted agents to manage them for us so we can gain more control over our digital lives.

When technologists see obstacles, they look for detours around them. Developer and author Gina Trapani says they hack their way out of restrictions. They sometimes flout the law because they know it is doomed to be two steps behind technology. But Lawrence Lessig, of Harvard's Berkman Center for Internet and Society, sees these questions in terms of law. Code is law, he says.[17] When Facebook's code says what is private and what is public by default, it creates statutes that govern the behavior of its users and communities. When Facebook finds its law in conflict with the norms and expectations of its community, one of them must change. The same is true even of the game company Zynga when it decides how its FarmVille agrarians will interact and how their economy of fake goods will operate. And certainly Google is aware that its code influences the behavior of both good actors and bad—the spammers who try to game its system. That is why Google publishes its software and design principles.[18] When Google did the work of China's censors, it was violating its number one principle, "Don't be evil."[19] Is the best frame of reference for the principles of publicness history, economics, property, hacking, code, statute, or constitutional law?

Inspired by Barlow's Declaration of Independence for Cyberspace, I was tempted to wish for the next logical step (for an American, at least): a

Constitution for Cyberspace. But deciding who has the right to negotiate for the net and compromising on a set of laws would be limiting and dangerous. It would be too easy for a new and centralized power structure to emerge to enforce that Constitution. Decentralization is what makes the internet the internet. And as Dewar cautions, we should not be too quick to structure the net before we know what it is and what it can accomplish.

Making the next logical leap, I had the hubris to draft a Bill of Rights in Cyberspace on my blog. I was hardly the first to try. In 2009, a group of Chinese intellectuals issued a set of principles covering freedom of speech, including the freedom of opinion and the right to anonymity.[20] The Association for Progressive Communications issued a charter starting in 2001 with a detailed set of rights. It was laden with many agendas and the tendency of committees to throw everything in with the kitchen sink.[21] The Internet Rights and Principles Coalition drafted a preliminary version of its bill of rights that was similarly broad. But then, collaboratively, the group did a good job distilling it down to ten key rights:[22] universality and equality; rights and social justice; accessibility; expression and association; privacy and data protection (including the right to use data encryption); life, liberty, and security; diversity; network equality; standards and regulation; and governance. The Brazilian Internet Steering Committee issued a more succinct set of Principles for the Governance and Use of the Internet.[23] A group of Facebook users issued their own social bill of rights that's quite specific to that community.[24] Some have written principles covering just government data.[25] These documents answer many needs with much good thinking.

What we need first, I think, is discussion. Through that we will begin to discern our shared principles for the internet and our public society. I doubt we will ever arrive at a single set of principles. But I am convinced that we must have debates around the notion of principles. In the course of that, some truths will become self-evident. We will come to examine what matters to us and what we must protect. We will expose different views, priorities, and needs. Most important, we will have an expression of some principles to point to when powerful institutions try to control our net and diminish our publicness, power, and freedom. "It is time for us to use the internet to save the internet," Al Franken told the entrepreneurs at South by Southwest.

I return one last time to Germany, bringing the discussion full circle. After reading my proposed principles on my blog, German Justice Minister Sabine Leutheusser-Schnarrenberger responded in the Frankfurter Allgemeine Zeitung. She argues at the start that "the potential of the digital world must not be strangled by anxious over-regulation." Net neutrality, she declares, is "the only guarantor of the free exchange of information." She calls for a charter for the internet built on democratic and ethical principles, including human rights and free speech. We agree. Then we disagree. She calls my philosophy "post-privacy" and says that a society's freedom is reflected in the protection given to a person's privacy, which she argues needs better safeguarding, updated for a new digital reality. "Personal information is not an abstract concept of ones and zeroes. It is the digital recording of a human being." We may not disagree as much as she thinks. In any case, we both see the need for this discussion about principles. "The digital world does not primarily need new laws, it needs universal digital values," she writes. "The internet community must intensify this process of discussion."[26] Right. This is precisely the kind of discussion we need. That is why I submit to you my set of principles—which address more than the internet and privacy—and ask you to share your perspective in an ongoing discussion at www.buzzmachine.com/publicparts.

Principles of Publicness

I. We have the right to connect.

If we cannot connect, we cannot speak. That is a new and necessary preamble to our First Amendment. Finland has declared internet access—high-speed at that—as a right of citizens. Whether countries should subsidize and provide access is a separate question. But once access is established, cutting it off should be seen as a violation of human rights. That's what a 2011 United Nations report said. "It's now a basic human right to have internet," Thomson Reuters CEO Tom Glocer told media executives in the Middle East. "Systematic denial of freedom of accessing information will lead to a revolution." [27]

II. **We have the right to speak.**

Freedom of speech is our cultural and legal default in the United States. That First Amendment protection should extend not just to information and opinions delivered by text but also to information delivered by applications and data. Yes, there need to be limitations—on child pornography online, for example. But beware the unintended consequences of attacking a specific problem with an overly broad response. To fight child porn, Australia proposed mandatory filters to block content—filters that could be used against any content. [28] We cannot manage everything to the worst case, to that which *might* offend someone, to that which *could* happen. We must not live by the lowest common denominator of fear and offense and the highest watermark of regulation, diminishing our most precious right of speech in the process.

III. **We have the right to assemble and to act.**

It is not enough to speak. Our tools of publicness enable us to organize, to gather together—virtually or physically—and to act as a group to demonstrate or to build.

IV. **Privacy is an ethic of knowing.**

We need protection of privacy. We also need to adapt our norms of privacy to new social tools and behaviors so we can better understand when something is said in confidence, when information should not be used without consent, what the harm is of spreading information, and how to give people more control of their information.

V. **Publicness is an ethic of sharing.**

The foundation of a more public society is the principle of sharing: recognizing the benefits of generosity, building tools that facilitate it, and protecting the product of it.

VI. Our institutions' information should be public by default, secret by necessity.

Openness is a better way to govern and a smarter way to do business.

VII. What is public is a public good.

When public information or the public space is diminished, the public loses. Secrecy too often serves the corrupt and tyrannical.

VIII. All bits are created equal.

When anyone gains the power to decide which bits, words, images, or ideas can or cannot pass freely through our network, it is no longer free.

IX. The internet must stay open and distributed.

"Let's give credit to the people who foresaw the internet, opened it up, designed it so it would not have significant choke points, and made it possible for random people including twenty-four-year-olds in a dorm to enter and create,"[29] says Eric Schmidt.

Before the 2011 meeting of the G8, French President Nicolas Sarkozy gathered a thousand leaders from government, technology, media, and universities in tents on the Tuileries Garden in Paris to consider governments' role in the future of the internet. I stood up at this first meeting of the e-G8 and welcomed the discussion of principles but then asked Sarkozy to take a Hippocratic oath for the net: "First, do no harm." Insulted, he mocked the question, contending that protecting intellectual property, security, privacy, and children would not harm. That's true, depending on how it's done—by enforcing laws that already exist or by trying to extend new authority over our new world. Sarkozy had talked of civilizing the internet and was portrayed in the French press as the sheriff riding into town to tame our Wild West. He argued that government must have a role in governing the net. "No one should forget," he declared, "that governments in our democracies are the only legitimate representatives of the populace." Well, tell that to the people of Tahrir Square. Their government was most certainly not their legitimate representative and they used the internet, their tool of publicness, to find their own voice. I left the meeting fearful of those who fear the internet and its change.[30]

I also left doubting the metaphor of internet-as-continent. Sarkozy—who'd rejected the notion of the net as a parallel universe—liked the phrase I used, "the eighth continent." Perhaps he saw himself planting his flag in its soil. The problem is, we don't leave our native lands to come to the net. We are all citizens of some nation, subject to its laws. Now we can also be citizens of this new society. This idea of the net as a new society is confounding to nations of politics and laws, for the net's own sovereignty depends on no one having sovereignty over it. That's how it was designed. That is its core principle and its disruptive power. The net, I've come to see, is a counterweight to the power and authority of government and of corporations. Is this at last the embodiment of Habermas' public sphere? I think of it more as a platform for the creation of publics. To be that, it must remain independent and free of those whose power it would check—that is, the people Sarkozy called to the Tuileries tents. If the net is our platform, it is up to us, the people of the net, to learn how to use it and to protect it, establishing and defending our own principles rather than waiting for companies to deliver theirs in terms of service or governments in laws.

As we face epochal change, it is fine and necessary to ask what could go wrong and to guard against our worst-case fears. But it is also vital that we recognize new opportunities, envisioning the sort of society we can build upon an ethic of sharing. We have the tools to do it. What Gutenberg's press brought to the early modern age, these tools now bring to anyone in this, the early digital age. They empower us. They grant us the ability to create, to connect, to organize, and to aggregate our knowledge. They provoke generosity and collaboration. They allow people to make their living in new ways and to build new industries and markets. They lower borders, even challenging our notion of nations. I hope that publicness serves to make us a more tolerant and trusting society. But this future is not assured. The choices are ours. The shape of our new world is up to us, the public.

Acknowledgments

I am indebted to many people who informed and influenced my thinking for this book, including particularly Paul Yachnin and the Making Publics project; Elizabeth Eisenstein; the Gutenberg Parenthesis project at the University of Southern Denmark; Jay Rosen, who taught me the importance of the idea of publics in relation to news and media; danah boyd, who rescued me from confusion on the topic of privacy; David Weinberger; Clay Shirky; Steven Berlin Johnson; and all those who generously granted me interviews and arranged them.

I delivered this manuscript to my editor, Ben Loehnen, not as a beta but as a pre-alpha. Through his patience and perseverance and his considerable talent for finding just the right threads to pull, he enabled it to become a book. I am most grateful. I thank my agent, Kate Lee of ICM, for her invaluable support and advice; she, too, requires patience with me. I'd also like to thank the team from Simon & Schuster: Jessica Abell, Kelly Welsh, Brian Ulicky, Sammy Perlmutter, Jon Karp, Marcella Berger, Richard Rhorer, Martha Schwartz and Michael Accordino. And I am grateful to Hollis Heimbouch at HarperCollins.

My wonderful family—my wife, Tammy, and children, Jake and Julia—suffer through my hours, moods, and distractions and give me the freedom that allows me to write. They also cope with the trials of living with a too-public husband and father. I cannot thank them enough for their love, forbearance, and inspiration.

I thank my colleagues at the City University of New York Graduate School of Journalism and the Tow-Knight Center for Entrepreneurial Journalism: Stephen Shepard, Judith Watson, Peter Hauck, Jeremy Caplan, Sandeep Junnarkar, Adam Glenn, John Smock, Jennifer McFadden, and the faculty, staff, and students of the school.

Notes

The notes and citations here will also appear online—at www.buzz machine.com/publicparts—whenever appropriate, as clickable links. There, I will add clickable links to most of the web sites mentioned in the book. From time to time, I plan to add further links and updates at the site. I cannot guarantee that all links will operate forever. The internet is unfortunately littered with dead links to dead ends.

At that site and on the Facebook page for this book—on.fb.me/public parts—I also urge you to join in the discussion about the ideas here. You can find me at my blog, BuzzMachine.com, and on Twitter @jeffjarvis (twitter.com/jeffjarvis).

Full references for titles given in short form will be found in the Bibliography.

Introduction: The Ages of Publicness

1. Solove, *Understanding Privacy,* p. 5
2. *The Origins of the Modern Public,* CBC, cbc.ca/ideas/episodes/fea tures/2010/04/26/the-origins-of-the-modern-public/
3. Warner, *Publics and Counterpublics,* p. 30
4. "SeeClickFix Integrates Data with San Francisco and Washington DC's New Open311 Systems," Future Cities Forum, futurecitiesforum.com/ content/seeclickfix-integrates-data-san-francisco-and-washington-dcs-new open311-systems
5. The phrase is inspired by New York University's Jay Rosen in the context of newly collaborative news and "the people formerly known as the audience." See Rosen's blog, June 27, 2006, archive.pressthink.org/2006/06/27/ ppl_frmr.html
6. "Bradley Cooper vs. Bradley Cooper . . . Sarah Ellison on Wall Street . . . ," Womens Wear Daily, wwd.com/media-news/fashion-memopad/bradley -cooper-vs-bradley-cooper-sarah-ellison-on-wall-street-3072929#/article/ media-news/fashion-memopad/bradley-cooper-vs-bradley-cooper-sarah -ellison-on-wallstreet-3072929?page=3
7. Doc Searls, "The Net: Free Infrastructure for Speech, Enterprise and Assembly," Harvard Law Blogs, January 13, 2010, blogs.law.harvard.edu/

doc/2010/01/13/the-net-free-infrastructure-for-speech-enterprise-and
-assembly/

8. Bernard Kouchner, "The Battle for the Internet," The New York Times,
   May 13, 2010, nytimes.com/2010/05/14/opinion/14iht-edkouchner.html

9. Peter Levin and Mehret Mandefro, draft of an article for Foreign Affairs

10. facebook.com/press/info.php?statistics

11. Sennett, The Fall of Public Man, p. 24

12. Note also the theory that biological evolution does not come in a steady
    march of small changes in our DNA brought on by natural selection but
    instead in rare and more drastic bursts of change as species split from each
    other. Wikipedia entry on punctuated equilibrium (en.wikipedia.org/wiki/
    Punctuated_equilibrium); Niles Eldredge and Stephen Jay Gould, "Punc-
    tuated Equilibria: An Alternative to Phyletic Gradualism," Speciation,
    1972 (blackwellpublishing.com/ridley/classictexts/eldredge.pdf); and Chris
    Venditti and Mark Pagel, "Speciation and Bursts of Evolution," Evolution:
    Education and Outreach 1 (2008): 274–80 (springerlink.com/content/
    vr7t4mx65j3x62w0/fulltext.pdf)

13. facebook.com/elshaheeed.co.uk

14. "Mission accomplished": After Mubarak and Bush, can we please retire the
    phrase?

15. Jackie Cohen, "Google's Wael Ghonim Thanks Facebook for Revolution,"
    All Facebook, February 11, 2011, allfacebook.com/googles-wael-ghonim
    -thanks-facebook-for-revolution-2011–02

16. Doc Searls, "Earthquake Turns TV Networks into Print," Harvard Law
    blogs, March 11, 2011, blogs.law.harvard.edu/doc/2011/03/11/earthquake
    -turns-tv-networks-into-print/

17. "Mubarak's Egypt NO MORE," Sandmonkey blog, February 12, 2011,
    sandmonkey.org/2011/02/12/mubaraks-egypt-no-more/

18. Warner, Publics and Counterpublics, p. 21

The Prophet of Publicness: Mark Zuckerberg

1. Disclosure: I was brought to Murdoch's retreat to lead a discussion with
   Zuckerberg and Gawker Media founder Nick Denton precisely because
   both men are so laconic

2. Norman Holland, "This Is Your Brain on Culture," Psychology Today, November 2, 2010, psychologytoday.com/blog/is-your-brain-culture/201011/
   the-social-network-aspergers-and-your-brain

3. Mark Harris, "Inventing Facebook," New York Magazine, September 17,
   2010, nymag.com/movies/features/68319/

4. Kirkpatrick, The Facebook Effect, pp. 189–92

5. Matt McKeon, "The Evolution of Privacy on Facebook," Matt McKeon
   blog, mattmckeon.com/facebook-privacy/

6. twitter.com/ajkeen/status/21592610173

7. "Grok," geekspeak coined by Robert Heinlein in Stranger in a Strange Land,

is defined in the *Oxford English Dictionary* as "to understand intuitively or by empathy," en.wikipedia.org/wiki/Grok

8. "Speech by Rupert Murdoch to the American Society of Newspaper Editors," April 13, 2005 (newscorp.com/news/news_247.html). Disclosure: I advised Murdoch's writer on the speech and earlier worked for TV Guide, then a division of their company, News Corporation. My last book was published by News Corporation's HarperCollins.

9. If I don't become your Facebook friend, please don't be insulted. I tend to use the service to organize and follow people I actually know. It also happens that my account has been broken for more than a year and I have been telling people that I have too many friends (5,000) when in reality I have far fewer. Frankly, I've found that to be convenient, sparing me from that awkward moment of rejecting someone who would be my friend, an act that feels rather like shoving a puppy. To handle just this situation, Facebook set up so-called fan pages, which eliminate the need to rule on friendships. Nonetheless, I find the idea of fan pages rather narcissistic; it's bad enough that we can all have fifteen minutes of fame, but to assume that we all have fans strikes me as just too Hollywood. So there is, instead, a fan page for this book at on.fb.me/publicparts

10. Zuckerberg's contention that we choose what to post on Facebook about ourselves is generally true. Except that your friends can talk about you and post pictures of you—though, at least on Facebook, you have control over whether you're named in the pictures.

11. Moore's law, formulated by Intel cofounder Gordon Moore, dictates that the number of transistors that can be placed on a computer chip doubles about every two years. That is the engine that has driven the exponential rate of technological change in society. See en.wikipedia.org/wiki/Moore's_law

12. Saul Hansell, "Zuckerberg's Law of Information Sharing," The New York Times blog, November 6, 2008, bits.blogs.nytimes.com/2008/11/06/zuckerbergs-law-of-information-sharing/

Public Choices

1. Sonia Phalnikar, ed., "Germany Threatens Google over Street View," Deutsche Welle, February 6, 2010, dw-world.de/dw/article/0,,5222701,00.html

2. Mat Greenfield, "Google Ghosts: New Street View Tech Could Wipe Humans from the Map," CNET, August 9, 2010, crave.cnet.co.uk/software/google-ghosts-new-street-view-tech-could-wipe-humans-from-the-map-50000258/?s_cid=33

3. Ulrich Schütz, "Leverkusen: SPD: Google soll 150 Euro pro Kilometer Zahlen," RP Online, July 23, 2010, rp-online.de/bergischesland/leverkusen/nachrichten/SPD-Google-soll-150-Euro-pro-Kilometer-zahlen_aid_885411.html (in German)

4. HF, "No Google Street View in Germany and Austria?," German Way blog, April 12, 2010, german-way.com/blog/2010/04/12/no-google-street-view-in-germany-and-austria/

5. youtube.com/watch?v=OMFBuHsKXb0

6. "Back Up Your Gmail: Google Threatens to End Email Service in Germany," Der Spiegel, June 25, 2007, spiegel.de/international/germany/0,1518,490492,00.html

7. Andreas Turk, "How Many German Households Have Opted-out of Street View?," Google Policy blog, October 21, 2010, googlepolicyeurope.blogspot.com/2010/10/how-many-german-households-have-opted.html

8. Though the site is in German, go to findedaspixel.de/spots/liste; at the bottom of the screen, you can navigate from one pixelated building to the next

9. Jeff Jarvis, "Germany, What Have You Done?," Buzzmachine blog, November 2, 2010 (www.buzzmachine.com/2010/11/02/germany-what-have-you-done/). In German: zeit.de/digital/internet/2010–11/street-view-jeff-jarvis-verpixelung

10. Cyrus Farivar, "Berlin Court Rules Google Street View Is Legal in Germany," Deutsche Welle, March 21, 2011, dw-world.de/dw/article/0,,14929074,00.html

11. Siobhán Dowling, "Google Knows More About Us than the KGB, or Gestapo," Der Spiegel, August 19, 2010, spiegel.de/international/germany/0,1518,712680,00.html

12. Ibid.

13. "Search Engine Barometer—May 2010," en.atinternet.com/Resources/Surveys/Search-Engine-Barometer/Search-Engine-Barometer-May-2010/index-1-2-6-199.aspx

14. "German court rejects lawsuit linked to Facebook 'Like' Button," Deutsche Welle, March 23, 2011, http://www.dw-world.de/dw/article/0,,14939422,00.html

15. "2 Swiss Pols, in Rare Move, Reveal Incomes, Assets," BusinessWeek, September 5, 2010, businessweek.com/ap/financialnews/D9I1O1000.htm

16. Jeff Jarvis, "Privacy, Publicness & Penises," Buzzmachine blog, April 22, 2010, www.buzzmachine.com/2010/04/22/privacy-publicness-penises/

17. Comment on Jeff Jarvis, "The German Privacy Paradox," Buzzmachine blog, February 11, 2010, www.buzzmachine.com/2010/02/11/the-german-privacy-paradox/#comment-408841

18. Eric Schmidt, "Letter to Google Inc. Chief Executive Officer," April 19, 2010, priv.gc.ca/media/nr-c/2010/let_100420_e.cfm

19. Jakof Adler, "Wie Ich mit Jeff Jarvis in der Sauna Saß," Jakuuub blog, April 14, 2010, jakuuub.de/2010/04/14/wie-ich-mit-jeff-jarvis-in-der-sauna-sass/ (in German)

20. Jeff Jarvis, "Cancer, the Sequel," Buzzmachine blog, January 21, 2011, www.buzzmachine.com/tag/prostate/

21. Jeff Jarvis, Buzzmachine blog, www.buzzmachine.com/tag/afib/

22. Jeff Jarvis, "The Small c and Me," Buzzmachine blog, August 10, 2009, www.buzzmachine.com/2009/08/10/the-small-c-and-me/

23. Jeff Jarvis, "Small c: The Penis Post," October 16, 2009, www.buzzmachine .com/2009/10/16/small-c-the-penis-post/

24. Tyndall Report, tyndallreport.com

25. Comment on Jeff Jarvis, "The Small c and Me," Buzzmachine blog, August 10, 2009, www.buzzmachine.com/2009/10/16/small-c-the-penis -post/#comment-403139

26. Comment on Jeff Jarvis, "The Small c and Me," Buzzmachine blog, August 10, 2009, www.buzzmachine.com/2009/10/16/small-c-the-penis -post/#comment-403546

27. The Wall Street Journal, "What They Know" series, wsj.com/wtk

28. Kai Biermann, "Betrayed by Our Own Data," Zeit Online, March 26, 2011, zeit.de/digital/datenschutz/2011–03/data-protection-malte-spitz

The Benefits of Publicness

1. Danny Sullivan, "Live Blogging Facebook CEO Mark Zuckerberg at Web 2.0 Summit," SearchEngineLand blog, November 16, 2010, searchengine land.com/live-blogging-ceo-mark-zuckerberg-at-web-20-56250

2. en.wikipedia.org/wiki/Theory_of_the_firm

3. "The New New Telco," January 7, 2011, Confused of Calcutta blog, con fusedofcalcutta.com/2011/01/07/the-new-new-telco/

4. Situationist, situationistapp.com

5. Dunbar number, en.wikipedia.org/wiki/Dunbar's_number

6. "IBM Is Committed to Linux and Open Source," March 2010, public.dhe .ibm.com/common/ssi/ecm/en/lxb03001usen/LXB03001USEN.PDF

7. "Finally, 1.0!," tcho.com/chocolate/story-of-one

8. Google flu trends, google.org/flutrends/about/how.html

9. Sun Valley–Google's Larry Page: Stop Stressing About Search Data Privacy," July 9, 2010, blogs.reuters.com/mediafile/2010/07/09/sun-valley-googles -larry-page-stop-stressing-about-search-data-privacy

10. Brian Stelter, "Tall Tales, Truth and My Twitter Diet," August 21, 2010, nytimes.com/2010/08/22/weekinreview/22stelter.html

11. twitter.com/brianstelter25

12. "The Power of Data, Zeitgeist Europe 2010," youtube.com/watch? v=10VXT0aGICc

13. heritagehealthprize.com

14. Jeff Jarvis, "Dell Learns to Listen," October 17, 2007, businessweek.com/ bwdaily/dnflash/content/oct2007/db20071017_277 576.htm

15. Howard Schultz, "Starbucks Chairman Warns of 'the Commoditization of the Starbucks Experience,'" February 23, 2007, starbucksgossip.typepad .com/_/2007/02/starbucks_chair_2.html

16. "One Last Post on the Super Bowl," February 10, 2011, groupon.com/blog/ cities/one-last-post-on-the-super-bowl/

17. Richard Sennett, *The Fall of Public Man,* p. 222
18. Solove, *Understanding Pivacy,* p. 95
19. Christopher Locke and David Weinberger in Rick Levine, Christopher Locke, Doc Searls, and David Weinberger, *The Cluetrain Manifesto,* p. 240
20. "Is the End of Privacy the End of Shame?," March 23, 2010, tweetagewaste land.com/2010/03/is-the-end-of-privacy-the-end-of-shame/
21. *This Week in Google,* twit.tv/twig
22. Arendt, *The Human Condition,* p. 58
23. Winer has often written about "future-safe archives". Two links: "Meeting at Library of Congress," November 3, 2010, scripting.com/stories/2010/11/03/meetingAtLibraryOfCongress.html and various articles indexed at google.com/search?q=site%3Ascripting.com+future+safe+archive
24. Jemima Kiss and Robbie Clutton, "Told You I Was Ill: All About Digital Death," March 14, 2011, jemimakiss.tumblr.com/post/4104890796/told -you-i-was-ill-all-about-digital-death
25. Andrew Sullivan, "The Daily Dish," The Atlantic, June 15, 2009
26. Clay Shirky, "The Net Advantage," Prospect 165, December 11, 2009
27. Evgeny Morozov, "Iran: Downside to the 'Twitter Revolution,'" Dissent 56, no. 4 (Fall 2009)
28. Shirky, "The Net Advantage"
29. Jeffrey Kofman and Ki Mae Huessner, "Libya's 'Love Revolution': Muslim Dating Site Seeds Protest," February 24, 2011, abcnews.go.com/Technology/muslim-dating-site-madawi-seeds-libyan-revolution/story?id=12981938&page=1
30. "FactCheck: How Many CCTV Cameras?," June 18, 2008, channel4.com/news/articles/society/factcheck+how+many+cctv+cameras/2291167
31. Cory Doctorow, "London Bus with 16 CCTV Cameras Inside," August 31, 2009, boingboing.net/2009/08/31/london-bus-with-16-c.html
32. James Risen and Eric Lichtblau, "Bush Lets U.S. Spy on Callers Without Courts," December 16, 2005, nytimes.com/2005/12/16/politics/16program.html
33. Andrew Levy, "Council Uses Spy Plane with Thermal Imaging Camera to Snoop on Homes Wasting Energy," March 24, 2009, dailymail.co.uk/news/article-1164091/Council-uses-spy-plane-thermal-imaging-camera-snoop -homes-wasting-energy.html
34. Haringey Interactive Heat Loss Map, seeit.co.uk/haringey/Map.cfm
35. "12-Year-Old Sued for Music Downloading," September 9, 2003, foxnews.com/story/0,2933,96797,00.html
36. Nate Anderson, "Obama 'IP Czar' Wants Felony Charges for Illegal Web Streaming," arstechnica.com/tech-policy/news/2011/03/obama-ip-czar -wants-felony-charges-for-illegal-web-streaming.ars
37. Ilse Aigner, "Wir passen den Datenschutz ans Internet-Zeitalter an," September 17, 2010, abendblatt.de/hamburg/article1634935/Wir-passen-den -Datenschutz-ans-Internet-Zeitalter-an.html
38. en.wikipedia.org/wiki/Umar_Farouk_Abdulmutallab

39. Adrian Chen, "Google Earth Used to Bust Illegal Swimming Pools," August 14, 2010, gawker.com/5612940/google-earth-our-newest-creepiest-crime+fighting-tool

40. Centers for Disease Control and Prevention,"Unintentional Drowning: Fact Sheet," cdc.gov/HomeandRecreationalSafety/Water-Safety/water injuries-factsheet.html

A History of the Private and the Public

1. kodak.com/global/en/corp/historyOfKodak/1878.jhtml
2. Solove, *Understanding Privacy,* p. 16
3. The New York Times archive, August 18, 1899
4. The New York Times, April 12, 1903
5. The New York Times, January 26, 1874
6. The New York Times, March 14, 1897
7. The New York Times, February 24, 1897
8. The New York Times, March 7, 1898
9. *Roberson v. Rochester Folding Box Co.,* 1902
10. The New York Times, July 2, 1902
11. The New York Times, April 12, 1903
12. Translation courtesy of my Twitter readers, seventy-five of whom, within minutes of my asking, gave me a variety of translations and pointed me to Wikipedia: en.wikipedia.org/wiki/De_minimis
13. The New York Times, July 2, 1902
14. The New York Times, August 23, 1902
15. The New York Times, July 27, 1904
16. Brandeis and Warren, "The Right to Privacy"
17. Amy Gajda, "What If Samuel D. Warren Hadn't Married a Senator's Daughter?: Uncovering the Press Coverage That Led to 'the Right of Privacy,'" Michigan State Law Review 35 (Spring 2008): 35–60, msulaw review.org/PDFS/2008-1/Gajda.pdf
18. Eisenstein, *The Printing Press as an Agent of Change,* p. 230
19. McKeon, *The Secret History of Domesticity,* p. 85, quoting John Robinson in *New Essays and Observations Divine and Morall* (London, 1628)
20. Ibid., pp. 54, 56
21. Westin, *Privacy and Freedom,* pp. 338, 52, 69–70, 3, 72–80, 86, 156, 160–162, respectively
22. Nissenbaum, *Privacy in Context,* p. 1
23. Schaar, *Das Ende der Privatsphäre,* pp. 5–8
24. Douglas Adams, The Sunday Times (London), August 29, 1999
25. Matt Ridley, "Humans: Why They Triumphed," The Wall Street Journal, May 22, 2010, http://onwsj.com/9769PX
26. Hind, *The Return of the Public,* p. 16
27. Habermas, *The Structural Transformation of the Public Sphere,* p. 11
28. Ibid., p. 6
29. Friedman, *Guarding Life's Dark Secrets,* p. 258

30. Sennett, *The Fall of Public Man,* p. 16

31. Habermas, *The Structural Transformation of the Public Sphere,* p. 11

32. Spacks, *Privacy,* pp. 1–2

33. Arendt, *The Human Condition,* p. 38

34. Warner, *Publics and Counterpublics,* p. 23

35. Spacks, *Privacy,* pp. 2, 5

36. Cited in Solove, *Understanding Privacy,* p. 81

37. Girouard, *Life in the English Country House,* p. 138

38. Spacks, *Privacy,* p. 7

39. Sennett, *The Fall of Public Man,* pp. 121–22

40. Solove, *Understanding Privacy,* p. 61

41. Spacks, *Privacy,* p. 3

42. makingpublics.mcgill.ca

43. Wilson and Yachnin, *Making Publics in Early Modern Europe,* pp. 1–2

44. www.cbc.ca/ideas/episodes/features/2010/04/26/the-origins-of-the-modern-public/

45. Peter Levin and Mehret Mandefro, draft of an article for Foreign Affairs

46. Marcus, "Cyberspace Renaissance," p. 389

47. For example: "The constitutional state predicated on civil rights pretended, on the basis of an effective public sphere, to be an organization of public power ensuring the latter's subordination to the needs of a private sphere itself taken to be neutralized as regards power and domination." Habermas, *The Structural Transformation of the Public Sphere,* p. 84

48. Though my German is poor, I will dare to question the translation of his title, as I believe the words in it carry much baggage in English. "Bürgerliche Gesellschaft" was translated as "bourgeois society" but perhaps that carries too much class distinction for American ears. Even though Habermas ends up describing just one class—that of the coffeehouses—I wonder whether "civic society" would be better. I had also wondered whether *Öffentlichkeit* could mean publicness, a state of being public, instead of the public sphere. But my friend Wolfgang Blau, the editor-in-chief of Zeit Online, explained to me that in German, *Öffentlichkeit* describes a space or a sphere, an abstract location or function in society rather than a mind-set or attitude.

49. Habermas, *The Structural Transformation of the Public Sphere,* p. 28

50. Cowan, *The Social Life of Coffee,* p. 90

51. Habermas, *The Structural Transformation of the Public Sphere,* p. 32

52. Cowan, *The Social Life of Coffee,* p. 87

53. Habermas, *The Structural Transformation of the Public Sphere,* pp. 25, 65, 70

54. Ibid., p. 1

55. Nancy Fraser, "Rethinking the Public Sphere," in *Habermas and the Public Sphere,* ed. Craig Calhoun, pp. 109–42

56. Warner, *Publics and Counterpublics,* p. 12

57. Habermas, *The Structural Transformation of the Public Sphere,* pp. 160–62

58. Ibid., pp. 2, 171, 147–49, 142, respectively

59. I said I'd spare you the sausage, but for those brave enough to explore notes, I can't resist sharing his fear that a "dialectic of a progressive 'societalization' of the state simultaneously with an increasing 'stateification' of society gradually destroyed the basis of the bourgeois public sphere." Ibid., p. 142

60. Ibid., p. 143

61. Habermas, "Political Communication in Media Society," a speech by Habermas before the International Communication Association published in Communication Theory 16 (2006): 423

62. Higgins, *The Raymond Williams Reader,* p. 46

63. Mills, *The Sociological Imagination,* quoted in Habermas, *The Structural Transformation of the Public Sphere,* p. 249

The Public Press

1. Eisenstein, *The Printing Press as an Agent of Change,* pp. 25, 36–37
2. Myron Gilmore, *The World of Humanism,* quoted in ibid., p. 28
3. McLuhan, *The Gutenberg Galaxy,* p. 90
4. Man, *The Gutenberg Revolution,* pp. 106–107
5. "The Invention of Paper," June 13, 2006, ipst.gatech.edu/amp/collection/museum_invention_paper.htm
6. Man, *The Gutenberg Revolution,* p. 14
7. Febvre and Martin, *The Coming of the Book,* p. 248
8. Eisenstein, *The Printing Press as an Agent of Change,* p. 33
9. Marcus, "Cyberspace Renaissance," p. 390
10. Eisenstein, *The Printing Press as an Agent of Change,* pp. 108, 81
11. E. P. Goldschmidt, *Gothic and Renaissance Bookbindings,* quoted in ibid., p. 49
12. Eisenstein, *The Printing Press as an Agent of Change,* p. 109
13. Lloyd A. Brown, *The Story of Maps,* quoted in ibid., p. 110
14. Gerald Strauss, *Sixteenth Century Encyclopedia* quoted in Eisenstein, *The Printing Press as an Agent of Change,* p. 109
15. openstreetmap.org
16. Eisenstein, *The Printing Press as an Agent of Change,* pp. 478, 110, 518
17. Eisenstein, "An Unacknowledged Revolution Revisited"
18. en.wikipedia.org/wiki/Pamela,_or_Virtue_Rewarded#Richardson.27s_revisions
19. Eisenstein, *Divine Art, Infernal Machine,* pp. 25, 55
20. Ann Blair, "Errata Lists and the Reader as Corrector," in Baron, Lindquist, and Shevlin, *Agent of Change,* p. 24
21. Eisenstein, *Divine Art, Infernal Machine,* p. 116
22. Ibid., p. 23
23. Eisenstein, *The Printing Press as an Agent of Change,* p. 76
24. McLuhan, *The Medium Is the Massage,* p. 50
25. Febvre and Martin, *The Coming of the Book,* p. 249
26. Pettegree, *The Book in the Renaissance,* p. 20

27. Ibid., pp. 44, 51
28. Eisenstein, *The Printing Press as an Agent of Change*, p. 229
29. Pettegree, *The Book in the Renaissance*, p. 65
30. McLuhan, *The Medium Is the Massage*, pp. 122, 50
31. McLuhan, *The Gutenberg Galaxy*, p. 250
32. Eisenstein, *The Printing Press as an Agent of Change*, pp. 72–75
33. Isaac Joubert, cited in ibid., pp. 244–45, 305
34. Eisenstein, *The Printing Press as an Agent of Change*, p. 432
35. Ibid., p. 66
36. George Sarton, *Six Wings*, cited in ibid., p. 506
37. Kapr, *Johann Gutenberg*, pp. 245–46
38. Eisenstein, *The Printing Press as an Agent of Change*, p. 363
39. Geoffrey Dickens, quoted in ibid., p. 303
40. Eisenstein, *The Printing Press as an Agent of Change*, p. 304
41. Eisenstein, *Divine Art, Infernal Machine*, p. 17
42. Munson and Warren, *James Carey: A Critical Reader*, p. 191
43. Ibid., p. 209
44. Ibid., p. 217
45. Ibid., p. 218
46. Sir John (Jack) Rankine Goody, *The Logic of Writing and the Organization of Society*, 1986, p. xi, cited in Jackaway, *Media at War*, p. 3
47. Jackaway, *Media at War*, pp. 7–8
48. Google response to Federal Trade Commission, 2010, p. 3, googlepub licpolicy.blogspot.com/2010/07/business-problems-need-business.html
49. Jackaway, *Media at War*, p. 8
50. M. G. Siegler, "Nicholas Negroponte: The Physical Book Is Dead in 5 Years," TechCrunch blog, August 6, 2010 (techcrunch.com/2010/08/06/physical-book-dead/). Here are Negroponte's predictions in 1984: "Nicholas Negroponte, in 1984, Makes 5 Predictions" (ted.com/talks/nicholas_negro ponte_in_1984_makes_5_predictions.html)
51. Nicholas Carr, "Is Google Making Us Stupid?," The Atlantic, July–August 2008, theatlantic.com/magazine/archive/2008/07/is-google-making-us-stu pid/6868/
52. Nicholas Carr, "Eric Schmidt's Second Thoughts," Rough Type blog, January 30, 2010, roughtype.com/archives/2010/01/eric_schmidts_s_1.php
53. John Naughton, "The Internet: Is It Changing the Way We Think?," The Observer, August 15, 2010, guardian.co.uk/technology/2010/aug/15/inter net-brain-neuroscience-debate
54. Marcus, "Cyberspace Renaissance," p. 396
55. The Gutenberg Parenthesis Research Forum, "The Gutenberg Parenthesis—Print, Book and Cognition" (sdu.dk/om_sdu/institutter_centre/ilkm/for skning/forskningsprojekter/gutenberg_projekt/positionpaper). See also this video with Professor Thomas Pettitt (niemanlab.org/2010/04/the-gutenberg-parenthesis-thomas-pettitt-on-parallels-between-the-pre-print-era-and-our-own-internet-age/). And this brief essay by Pettitt: "Before the Gutenberg

Parenthesis: Elizabethan-American Compatibilities" (web.mit.edu/comm
-forum/mit5/papers/pettitt_plenary_gutenberg.pdf).

56. McLuhan, *The Medium Is the Massage,* pp. 44–45

57. Marcus, "Cyberspace Renaissance," p. 401

What Is Privacy?

1. Solove, *Understanding Privacy,* p. 7

2. Judith Jarvis Thomson, "The Right to Privacy," in *Philosophical Dimensions of Privacy: An Anthology,* 1984, p. 272, cited in ibid., p. 7

3. Lillian R. BeVier, "Information About Individuals in the Hands of Government: Some Reflections on Privacy Protection," William and Mary Bill of Rights Journal, 1995, p. 458, cited in Solove, *Understanding Privacy,* p. 7

4. Raymond Wacks, *Law, Morality, and the Private Domain,* 2000, p. 222, cited in Solove, *Understanding Privacy,* p. 45

5. Westin, *Privacy and Freedom,* p. 7

6. William L. Prosser, "Privacy," California Law Review 48, no. 3 (August 1960): 389, californialawreview.org/assets/pdfs/misc/prosser_privacy.pdf

7. Solove, *Understanding Privacy,* p. 103

8. Brandeis and Warren, "The Right to Privacy"

9. Westin, *Privacy and Freedom,* p. 348

10. Ibid., p. 33

11. Lane, *American Privacy,* pp. 153–54

12. The Washington Post and Time, cited in ibid., pp. 141–42

13. *Olmstead v. United States,* 1928, p. 277, cited in Solove, *Understanding Privacy,* p. 17

14. Lane, *American Privacy,* pp. 153–54

15. Spacks, *Privacy,* p. 3

16. Ibid., pp. 3, 7, 140, 7, 88, 141, 13, 14, 15, respectively

17. McKeon, *The Secret History of Domesticity,* p. 61

18. Solove, *Understanding Privacy,* pp. 26–27, 31–32, 25, respectively

19. Ibid., p. 33

20. Ibid., pp. 74–75

21. Nissenbaum, *Privacy in Context,* p. 127

22. Westin, *Privacy and Freedom,* pp. 34, 35

23. Ibid., p. 331

24. Jamie Court, "Will Google Maps New Street View Tricycles Take Pictures of Our Kids' Playground?," The Huffington Post, March 2, 2011, huffing tonpost.com/jamie-court/will-google-maps-new-stre_b_830430.html

25. Simson Garfinkel and Beth Rosenberg, "Face Recognition: Clever or Just Plain Creepy?," Technology Review, technologyreview.com/comput ing/22234/?a=f

26. Greg Sterling, "Privacy, 'The Creepy Line' and Beyond: It's Not Just about Google," Search Engine Land blog, October 8, 2010, search engineland.com/privacy-the-creepy-line-and-beyond-its-not-just-about-google -52563

27. danah boyd, "Making Sense of Privacy and Publicity," March 13, 2010, danah.org/papers/talks/2010/SXSW2010.html

28. Polly Sprenger, "Sun on Privacy: 'Get over It,'" Wired.com, January 26, 1999, wired.com/politics/law/news/1999/01/17538

29. Amanda Lenhart and Mary Madden, "Teens, Privacy and Online Social Networks," Pew Internet, April 18, 2007, pewinternet.org/Reports/2007/Teens-Privacy-and-Online-Social-Networks/1-Summary-of-Findings.aspx

30. danah boyd, "Favorite Myth-making News Articles?," September 11, 2010, zephoria.org/thoughts/archives/2010/09/11/favorite-myth-making-news-articles.html

31. William McGeveran, "Finnish Employers Cannot Google Applicants," Harvard Law blogs, November 15, 2006, blogs.law.harvard.edu/infolaw/2006/11/15/finnish-employers-cannot-google-applicants/

32. "Saving Jobseekers from Themselves: New Law to Stop Companies from Checking Facebook Pages in Germany," Der Spiegel, August 23, 2010, spiegel.de/international/germany/0,1518,713240,00.html

33. Craig Schneider, "Gwinnett Hospital Won't Hire Smokers," The Atlanta Journal-Constitution, July 6, 2010, ajc.com/health/gwinnett-hospital-wont-hire-565480.html

34. Prosser, "Privacy," p. 391

35. "FTC Charges Deceptive Privacy Practices in Google's Rollout of Its Buzz Social Network," March 30, 2011, ftc.gov/opa/2011/03/google.shtm

36. Children's Online Privacy Protection Act, coppa.org/coppa.htm

37. "What We're Driving At," Google official blog, October 9, 2010, googleblog.blogspot.com/2010/10/what-were-driving-at.html

38. "Zelfrijdende Auto's, Eindelijk!," October 18, 2010, auto-en-vervoer.infonu.nl/auto/62126-zelfrijdende-autos-eindelijk.html

39. Sebastian, "Das Google Auto," October 11, 2010, kritikfabrik.de/2010/10/das-google-auto/

40. "Lessons from Online Bullying," CBS News, October 1, 2010, cbsnews.com/video/watch/?id=6919852n&tag=related;photovideo

How Public Are We?

1. http://www.facebook.com/press/info.php?statistics

2. Nicholas Jackson, "It's Official: Most American Adults Are Using Facebook," The Atlantic, March 25, 2011 (theatlantic.com/technology/archive/2011/03/its-official-most-americans-adults-are-using-facebook/73023/). Study by Edison Research: Tom Webster, "Facebook Achieves Majority," March 24, 2011 (edisonresearch.com/home/archives/2011/03/facebook_achieves_majority.php)

3. Amanda Lenhart, Kristen Purcell, Aaron Smith, and Kathryn Zickuhr, "Social Media and Young Adults," February 3, 2010, Pew Internet & American Life Project, pewinternet.org/Reports/2010/Social-Media-and-Young-Adults.aspx

4. twitter.com/about

5. "Internet 2010 in Numbers," January 12, 2011, royal.pingdom.com/2011/01/12/internet-2010-in-numbers/

6. Lenhart et al., "Social Media and Young Adults"

7. Bit.ly, which is used to shorten web addresses so they can fit into tweets, passed 4 billion clicks in early 2010, 8 billion in 2011. It accounts for only some of Twitter's traffic.

8. Alexia Tsotsis, "Flickr Hits Its 5 Billionth Photo, and Here It Is," TechCrunch blog, September 18, 2010, techcrunch.com/2010/09/18/flickr-5-billionth-photo/

9. "Internet 2010 in Numbers," http://royalpingdom.com/2011/01/12/internet-2010-in-numbers/

10. Disclosure: I am an investor in Covestor.

11. "Eric Schmidt: Every 2 Days We Create as Much Information as We Did Up to 2003," TechCrunch blog, August 4, 2010, http://techcrunch.com/2010/08/04/schmidt-data/

12. From Pew Internet & American Life Project surveys, including Reputation Management and Social Media, 2010; The Future of Social Relations, 2010; The Social Side of the Internet, 2010; Social Media & Mobile Internet Use Among Teens and Young Adults, 2010; Millennials Will Make Online Sharing in Networks a Lifelong Habit, 2010; Social Isolation and New Technology, 2009; Neighbors Online, 2010

13. Jeff Jarvis, Picks & Pans, People, September 5, 1988

14. Winfrey deleted her video, thus I have no link.

15. en.wikipedia.org/wiki/Justin.tv

16. blogs.wsj.com/wtk/

17. Mark Baard, "RFID: Sign of the (End) Times?," Wired.com, June 6, 2006, wired.com/science/discoveries/news/2006/06/70308

18. Miguel Bustillo, "Wal-Mart Radio Tags to Track Clothing," The Wall Street Journal, July 23, 2010, http://on.wsj.com/bzsfig

19. "PGP-10," personalgenomes.org/pgp10.html

20. Emily Gould, "Exposed," The New York Times Magazine, May 25, 2008, nytimes.com/2008/05/25/magazine/25internet-t.html?ref=emilygould

21. Steven Berlin Johnson, "In Praise of Oversharing," Time, May 20, 2010, time.com/time/business/article/0,8599,1990586,00.html

22. Mark Dery, "Have We No Sense of Decency Sir, at Long Last?: On Adult Diapers, Erectile Dysfunction, and Other Joys of Oversharing," True/Slant, June 7, 2010, trueslant.com/markdery/2010/06/07/have-we-no-sense-of-decency-sir-at-long-last-on-adult-diapers-erectile-dysfunction-and-other-joys-of-oversharing/

23. Gould, "Exposed"

24. The site has been taken down.

The Public You

1. cnbc.com/id/33831099?__source=vty|insidegoogle|&par=vty

2. Holman W. Jenkins Jr., "Google and the Search for the Future," The Wall Street Journal, August 14, 2010, http://on.wsj.com/aippTA

3. Judith Timson, "The Net Killed Sexual Privacy," September 2, 2010, theglobeandmail.com/life/family-and-relationships/the-net-killed-sexual -privacy/article1694613/

4. World of Warcraft help thanks to twitter friends: twitter.com/sikemapleton/ status/22461126238

5. Jeffrey Rosen, "The Web Means the End of Forgetting," July 21, 2010, ny times.com/2010/07/25/magazine/25privacy-t2.html

6. Scott Rosenberg, "Does the Web Remember Too Much—or Too Little?," July 26, 2010, wordyard.com/2010/07/26/the-end-of-forgetting-and-the -danger-of-forgetting/

7. Comment by David Weinberger at Buzzmachine, www.buzzmachine .com/2010/08/05/the-price-of-privacy/#comment-422091

8. Jane Hoffman, "A Girl's Nude Photo, and Altered Lives," The New York Times, March 26, 2011, reporting on an unfortunate case of "sexting"—a girl sending a naked picture to herself, which she sent to another girl, which then spread out of control. The piece is reasoned and not alarmist. It soberly and intelligently examines the issues and challenges of dealing with such a case. Yes, this is an edge case but one all parents want their children to avoid (nytimes.com/2011/03/27/us/27sexting.html).

9. Jeff Jarvis, "PDF: Eric Schmidt," Buzzmachine blog, May 18, 2007, www .buzzmachine.com/2007/05/18/pdg-eric-schmidt/

10. Holman W. Jenkins, Jr., "Google and the Search for the Future," The Wall Street Journal, http://onwsj.com/aippTA

11. Jeff Jarvis: "Dell Learns to Listen," BusinessWeek, October 17, 2007, businessweek.com/bwdaily/dnflash/content/oct2007/db20071017_277576 .htm

12. Facebook Help Center, facebook.com/help/?page=408#!/help/?faq=16162

13. Katrine Bussey, "Labour Accused as Candidate Sacked over 'Offensive' Tweets," The Independent, April 9, 2010, independent.co.uk/news/uk/poli tics/labour-accused-as-candidate-sacked-over-offensive-tweets-1940351.html

14. Jim Edwards, "Worst Twitter Post Ever: Ketchum Exec Insults Fedex Client on Mini-blog," BNet, January 20, 2009, bnet.com/blog/advertising-busi ness/worst-twitter-post-ever-ketchum-exec-insults-fedex-client-on-mini -blog/256

15. Andy Alexander, "Post Columnist Mike Wise Suspended for Fake Twitter Report," The Washington Post, voices.washingtonpost.com/ombudsman -blog/2010/08/post_columnist_mike_wise_suspe.html

16. Hutch Carpenter, "How to Tweet Your Way Out of a Job," March 17, 2009, bhc3.wordpress.com/2009/03/17/how-to-tweet-your-way-out-of-a-job/

17. Jameson Cook, "Facebook Post May Get Juror in Trouble," The Oakland Press, August 29, 2010, theoaklandpress.com/articles/2010/08/29/news/ cops_and_courts/doc4c7b104667b75071959419.txt

The Sharing Industry

1. Sam Lessin, speaking at the NY Tech Meetup, August 2010, livestream .com/nytechmeetup/video?clipId=pla_dae43c2e-9cf4-4cea-8bd7-5f0f9ee 32c52&utm_source=lslibrary&utm_medium=ui-thumb

2. twitter.com/lisa_dawson/status/21596110529

3. en.wikipedia.org/wiki/Blog

4. If you're too young to remember, see "The *Exciting* History of Carbon Paper!" (kevinlaurence.net/essays/cc.php). Carbon paper's first recorded use was in 1806.

5. Owen, *Copies in Seconds*

6. motion.kodak.com/US/en/motion/Products/Production/Spotlight_on_Super_8/ Super_8mm_History/index.htm

7. en.wikipedia.org/wiki/Mimeograph

8. en.wikipedia.org/wiki/Tripod.com

9. Bakewell, *How to Live,* pp. 25, 5, 9, respectively

10. Alan Rusbridger, "Why Twitter Matters for Media Organisations," November 19, 2010, guardian.co.uk/media/2010/nov/19/alan-rusbridger-twitter

11. "Nick Kristof Explains How Twitter Is Preventing Atrocities in the Middle East" (video), tvsquad.com/2011/03/30/piers-morgan-tonight-twitter-helps -prevent-atrocities-in-the/

12. twitter.com/delloutlet

13. Marshall Kirkpatrick, "The First Hashtag Ever Tweeted on Twitter—They Sure Have Come a Long Way," February 4, 2011, readwriteweb.com/ar chives/the_first_hashtag_ever_tweeted_on_twitter_-_they_s.php

14. Jessica Guynn and John Horn, "Twitter: A New Box-Office Miracle?," April 20, 2010, articles.latimes.com/2010/apr/02/business/la-fi-ct-twitter3 -2010apr03

15. Johan Bollen, Huina Mao, and Xiao-Jun Zeng, "Twitter Mood Predicts the Stock Market," Journal of Computational Science 2, no. 1 (March 2011): 1–8, arxiv.org/abs/1010.3003

16. Kate Kelly, "Hedge Fund to Predict Markets Using Twitter," March 8, 2011, cnbc.com/id/41948279

17. Tim Burton's Cadavre Exquis, burtonstory.com/connect.php

18. Crowley's Flickr, flickr.com/photos/dpstyles

19. twitter.com/blippy/statuses/15504284300877824

20. en.wikipedia.org/wiki/Signalling_(economics)

21. TripAdvisor fact sheet, tripadvisor.com/PressCenter-c4-Fact_Sheet.html

22. You have to see it to believe it. So go here to find the results of the in- terestingness algorithm and keep clicking for more: flickr.com/explore/ interesting/7days/

23. en.wikipedia.org/wiki/Josh_Harris_(internet)

24. Disclosure: I have worked with Samson.

25. "Sen. John D. Rockefeller IV Holds a Hearing on Consumer Online Pri- vacy, Panel 1," findarticles.com/p/news-articles/political-transcript-wire/ mi_8167/is_20100727/sen-john-rockefeller-iv-holds/ai_n54618189/

The Radically Public Company

1. Web 2.0 Summit 2010, "A Conversation with Mark Zuckerberg," youtube
   .com/watch?v=Czw-dtTP6oU

2. Ilya Marritz, "Regulators Push for Disclosure on Household Cleaner Ingredients," March 3, 2011, wnyc.org/articles/wnyc-news/2011/mar/03/ilya
   -add-text/

3. Here is Vaidhyanathan on the question of whether Google is evil in a debate: youtube.com/watch?v=lszQ-lO4Tk4; here is my response: youtube
   .com/watch?v=YQWGIEKY_sg&feature=player_embedded

4. simpsons.wikia.com/wiki/The_Homer

5. Sam Abuelsamid, "Jeff Jarvis Suggests Detroit Should Build a Beta Google-
   mobile," January 31, 2009, autoblog.com/2009/01/31/jeff-jarvis-suggests
   -detroit-should-build-a-beta-googlemobile/

6. "Toyota 'Unintended Acceleration' Has Killed 89," May 25, 2010, cbsnews
   .com/stories/2010/05/25/business/main6518794.shtml

7. Publishers get 68 percent of AdSense ads on content sites and 51 percent
   of search ads. See "The AdSense Revenue Share," May 24, 2010, adsense
   .blogspot.com/2010/05/adsense-revenue-share.html

8. code.google.com/chromium/ and sites.google.com/a/chromium.org/dev/
   chromium-os

9. Jeff Jarvis, "Google News," January 29, 2010, www.buzzmachine
   .com/2010/01/29/google-news-2/

10. Eric Schmidt, "Preparing for the Big Mobile Revolution," Harvard Business
    Review, hbr.org/web/extras/hbr-agenda-2011/eric-schmidt

11. John Paxton's Digital First blog, jxpaton.wordpress.com

12. See the rest of his tweets from the talk at "Create Stories Using Social
    Media," storify.com/jeffjarvis/jxpatons-twitter-lecture-to-newspapers

13. Disclosure: I write and have consulted for The Guardian.

14. Robert Booth, "Trafigura: A Few Tweets and Freedom of Speech Is Re-
    stored," October 13, 2009, guardian.co.uk/media/2009/oct/13/trafigura
    -tweets-freedowm-of-speech?INTCMP=SRCH

15. Alan Rusbridger, "The Hugh Cudlipp Lecture: Does Journalism Exist?,"
    January 25, 2010, guardian.co.uk/media/2010/jan/25/cudlipp-lecture-alan
    -rusbridger

16. Disclosures: This book was to be published by HarperCollins. When I saw
    that I was going to be critical of its parent, News Corporation, I decided to
    move to another publisher, Simon & Schuster. I gave advice to Murdoch's
    speechwriter for the talk quoted in this section. I worked for TV Guide,
    then part of News Corporation, for six years, from 1991, and then for
    Delphi, an internet company News Corporation had bought. I was paid to
    speak at a News Corporation retreat. Regarding The Guardian: I have writ-
    ten about new media for the paper and been brought in to speak to staff.

17. David Kravets, "Murdoch Calls Google, Yahoo Copyright Thieves—Is He
    Right?," April 3, 2009, wired.com/threatlevel/2009/04/murdoch-says-go/

18. "Speech by Rupert Murdoch to the American Society of Newspaper Editors," April 13, 2005, newscorp.com/news/news_247.html

19. John Lippman, "Murdoch Set to Buy Delphi Data Services," Los Angeles Times, September 2, 1993, articles.latimes.com/1993-09-02/business/fi-30898_1_electronic-newspaper

20. Arianna Huffington, "Journalism 2009: Desperate Metaphors, Desperate Revenue Models, and the Desperate Need for Better Journalism," December 1, 2009, huffingtonpost.com/arianna-huffington/journalism-2009-desperate_b_374642.html

21. Rafat Ali, "World Newspaper Congress: Dow Jones CEO: Beware of Geeks Bearing Gifts," December 1, 2009, paidcontent.org/article/419-world-newspaper-congress-dow-jones-ceo-beware-of-geeks-bearing-gifts/

22. "Comments on Federal Trade Commission's News Media Workshop and Staff Discussion Draft on 'Potential Policy Recommendations to Support the Reinvention of Journalism,'" July 20, 2010, docs.google.com/viewer?url=http%3A%2F%2Fgoogle.com%2Fgoogleblogs%2Fpdfs%2Fgoogle_ftc_news_media_comments.pdf

23. online.wsj.com/wtk

24. Seth Godin, "Moving On," August 23, 2010, sethgodin.typepad.com/seths_blog/2010/08/moving-on.html

25. Seth Godin, "The Domino Project," December 8, 2010, sethgodin.typepad.com/seths_blog/2010/12/the-domino-project.html

26. publicparts@buzzmachine.com

27. Best Buy Twelpforce Twitter account, twitter.com/twelpforce

28. Steve Myers, "How 'PriceOfWeed' Uses the Crowd to Fill an Illicit Information Gap," September 28, 2010, poynter.org/column.asp?id=136&aid=191657

29. "On a Bet, Party People Fill KLM Flight to Miami Using Twitter," December 28, 2010, springwise.com/tourism_travel/fly2miami/

30. en.wikipedia.org/wiki/Open-book_management

31. John Case, *Open-Book Management: The Coming Business Revolution,* amazon.com/Open-Book-Management-Coming-Business-Revolution/dp/0887308023

32. Here is a list of open-book management articles from Inc.: inc.com/guides/hr/23178.html

33. Jim McElgunn, "Open-Book Management: Doing Business in the Buff," Profit, May 2010

34. strategy.wikimedia.org/wiki/Main_Page

35. Jeff Jarvis, "Wiki Life," August 31, 2010, www.buzzmachine.com/2010/08/31/wiki/

36. Jay Rosen, "Asa Dotzler, 'Community Guy' at Mozilla Foundation, Talks to NewAssignment.Net," November 1, 2006, newassignment.net/blog/jay_rosen/so_i_went_to_see_asa_doztler_the_community_guy_at_mozilla_foundation

37. Scott Thomas, "Designing Obama," kickstarter.com/projects/simplescott/designing-obama

38. designing-obama.com

39. Dan Provost and Tom Gerhardt, "Glif—iPhone 4 Tripod Mount & Stand," kickstarter.com/projects/danprovost/glif-iphone-4-tripod-mount-and-stand?ref=live

40. Scott Wilson, "TikTok + LunaTik Multi-Touch Watch Kits," kickstarter.com/projects/1104350651/tiktok-lunatik-multi-touch-watch-kits

41. MIT Open Courseware, ocw.mit.edu

42. Leisa Anslinger and Daniel S. Mulhall, "Building a Google-y Church," Church, Fall 2009, churchmagazine.org/issue/0909/upf_building_a_google.php

By the People

1. Sifry, *WikiLeaks and the Age of Transparency,* p. 21

2. collateralmurder.com

3. Bruce Schneier, "Schneier on Security," schneier.com/blog/archives/2010/12/wikileaks_1.html

4. Sifry, *WikiLeaks and the Age of Transparency,* p. 163

5. Jeff Jarvis, "Just Saying," December 7, 2010, www.buzzmachine.com/2010/12/07/just-saying/

6. Sifry, *WikiLeaks and the Age of Transparency,* pp. 76–77

7. Lawrence Lessig, "Against Transparency: The Perils of Openness in Government," The New Republic, October 9, 2009, tnr.com/article/books-and-arts/against-transparency

8. Ibid.

9. Vivek Wadhwa, "The Goldmine of Opportunities in Gov. 2.0," October 23, 2010, techcrunch.com/2010/10/23/the-goldmine-of-opportunities-in-gov-2-0/

10. Tim O'Reilly, "Government as a Platform," in Lathrop and Ruma, *Open Government,* p. 12

11. Watch an inspiring video of Malamud at the Gov 2.0 conference here: youtube.com/watch?v=KemM5-s-bLU

12. public.resource.org/sec.gov/

13. Nathan Halverson, "Sebastopol Digital Innovator Wins $2 Million Google Grant," September 24, 2010, pressdemocrat.com/article/20100924/articles/100929642?p=1&tc=pg

14. David G. Robinson, Harlan Yu, William P. Zeller, and Edward W. Felten, "Government Data and the Invisible Hand," Yale Journal of Law & Technology 11 (2009):160, papers.ssrn.com/sol3/papers.cfm?abstract_id=1138083

15. Data.gov and other transparency projects were threatened with budget cutbacks and termination in 2011. See Daniel Schuman, "Budget Technopocalypse: Proposed Congressional Budgets Slash Funding for Data Transparency," March 23, 2011, sunlightfoundation.com/blog/2011/03/23/

transparency-technopocalypse-proposed-congressional-budgets-slash-fund
ing-for-data-transparency/

16. Simon Rogers, "My Top Ten data.gov.uk Datasets," October 7, 2010, data
.gov.uk/blog/my-top-ten-datagovuk-datasets-guest-post-simon-rogers

17. "The Framework," nationalarchives.gov.uk/information-management/gov
ernment-licensing/the-framework.htm

18. Carl Malamud, "Oregon: Our Laws Are Copyrighted and You Can't Publish
Them," April 15, 2008, boingboing.net/2008/04/15/oregon-our-laws-are.html

19. "NYC's Urban Forest: The 2006 Street Tree Census," milliontreesnyc.org/
html/urban_forest/urban_forest_census.shtml

20. U.S. Government Printing Office, "Federal Register 2.0," youtube.com/
watch?v=ADhP0KSmjkQ

21. McKeon, The Secret History of Domesticity, p. 57

22. Michael Schudson, "Was There Ever a Public Sphere? If So, When? Reflections on the American Case," in Calhoun, Habermas and the Public Sphere,
p. 155

23. Westin, Privacy and Freedom, p. 45

24. David Ferriero, "Federal Register 2.0," July 26, 2010, whitehouse.gov/
blog/2010/07/26/federal-register-20

25. Facebook translations, facebook.com/apps/application.php?id=4329892722
&v=info

26. Ushahidi, "2010 Earthquake in Haiti," haiti.ushahidi.com

27. Eliot Van Buskirk, "Snowmageddon Site Crowdsources Blizzard Cleanup,"
February 10, 2010, wired.com/epicenter/2010/02/snowmageddon-crowd
-sources-blizzard-clean-up/

28. The Ushahidi platform ushahidi.com/products/ushahidi-platform

29. en.wikipedia.org/wiki/Pirate_Parties_International

30. Sifry, WikiLeaks and the Age of Transparency, p. 56

31. Anthony Painter, "A Clash of Networks and Institutions," March 16, 2011,
labourlist.org/anthony-painter-a-clash-of-networks-and-institutions

32. Evgeny Morozov, "Picking a Fight with Clay Shirky," January 15, 2011,
neteffect.foreignpolicy.com/posts/2011/01/15/picking_a_fight_with_clay_
shirky

33. Malcolm Gladwell, "Does Egypt Need Twitter?," February 2, 2011, new
yorker.com/online/blogs/newsdesk/2011/02/does-egypt-need-twitter.html

34. Comment on Mathew Ingram, "Was What Happened in Tunisia a Twitter Revolution?," January 14, 2011, gigaom.com/2011/01/14/was-what
-happened-in-tunisia-a-twitter-revolution/#comment-575765

35. Adrian Johns, "How to Acknowledge a Revolution," The American Historical Review 107, no. 1 (February 2002). Eisenstein responds to Johns'
critique of her in The Nature of the Book: Print and Knowledge in the Making
(Chicago: University of Chicago Press, 1998) here: historycooperative
.org/journals/ahr/107.1/ah0102000087.html; Johns responds here: history
cooperative.org/journals/ahr/107.1/ah0102000106.html

36. John Naughton, "The Internet: Everything You Ever Need to Know," June 20, 2010, guardian.co.uk/technology/2010/jun/20/internet-everything -need-to-know

The New World

1. "Latest Directives from the Ministry of Truth," China Digital Times, chinadigitaltimes.net/2011/03/latest-directives-from-the-ministry-of-truth -march-10-18-2011/

2. Dan Sabbagh, "WikiLeaks Cables Blame Chinese Government for Google Hacking," December 4, 2010, guardian.co.uk/technology/2010/dec/04/ wikileaks-cables-google-china-hacking

3. investor.google.com/corporate/code-of-conduct.html

4. Jeff Jarvis, "Eric Schmidt on the New World," July 2, 2009, www.buzzma chine.com/2009/07/02/eric-schmidt-on-the-new-world/

5. James Dewar, "The Information Age and the Printing Press: Looking Backward to See Ahead," 1998, rand.org/pubs/papers/P8014/index2.html

6. Alexia Tsotsis, "Flickr Confirms Taking Down Egyptian Blogger's photos, Cites Community Guidelines Violation," March 11, 2011, techcrunch .com/2011/03/11/flickr/

7. "Flickr Censors Egyptian State Security Police Photos," March 13, 2011, lossofprivacy.com/index.php/2011/03/flickr-censors-egyptian-state-security -police-photos/

8. Dewar, "The Information Age and the Printing Press"

9. Viviane Reding, "Your Data, Your Rights: Safeguarding Your Privacy in a Connected World," March 16, 2011, europa.eu/rapid/pressReleasesAction.do ?reference=SPEECH/11/183&format=HTML&aged=0&language=EN&gui Language=en

10. twitter.com/privacylawyer/status/49513438417928192

11. Hillary Rodham Clinton, "Remarks on Internet Freedom," January 21, 2010, state.gov/secretary/rm/2010/01/135519.htm

12. Gerri Peev, "How Hillary Clinton Ordered U.S. Diplomats to Spy on UN Leaders," November 29, 2010, dailymail.co.uk/news/article-1333920/ WikiLeaks-Hillary-Clinton-ordered-U-S-diplomats-spy-UN-leaders.html

13. en.wikipedia.org/wiki/Sullivan_Principles

14. John Perry Barlow, "A Declaration of the Independence of Cyberspace," projects.eff.org/~barlow/Declaration-Final.html

15. Ibid.

16. Marc Davis, "Your Digital Life Data Is Bankable Currency," September 1, 2010, networkworld.com/community/blog/microsofts-davis-privacy-your -digital-life-da

17. See his books *Code and Other Laws of Cyberspace* (New York: Basic Books, 1999) and the update of that, *Code: Version 2.0* (New York: Basic Books, 2006)

18. Google, "Software principles," google.com/corporate/software_principles .html and Google, "Google user experience," google.com/corporate/ux.html

19. Google, "Our philosophy," google.com/corporate/tenthings.html
20. Ling Canzhou et al., "Internet Human Rights Declaration," October 8, 2009, underthejacaranda.wordpress.com/2009/10/08/internet-human-rights-declaration/
21. "APC Internet Rights Charter," apc.org/en/node/5677
22. "Internet Rights and Principles Coalition," irpcharter.org
23. Brazilian Internet Steering Committee, "Principles for the governance and use of the internet," cgi.br/english/regulations/resolution2009-003.htm
24. facebook.com/CFPBillOfRights
25. "Open Government Data Principles," resource.org/8_principles.html
26. Sabine Leutheusser-Schnarrenberger, "Das Recht auf Vernetzung," April 25, 2011, Frankfurter Allgemeine Zeitung, faz.net/artikel/C30833/digitale-sicherheit-das-recht-auf-vernetzung-30335119.html
27. Bindu Suresh Rai, "Social Media Revolution Unstoppable: Glocer," March 18, 2011, emirates247.com/business/technology/social-media-revolution-unstoppable-glocer-2011-03-18-1.369832
28. Tanalee Smith, "Australia Delays Internet Filter to Review Content," July 9, 2010, abcnews.go.com/Technology/wireStory?id=11123037
29. Jeff Jarvis, "Eric Schmidt on the New World," July 2, 2009, www.buzzmachine.com/2009/07/02/eric-schmidt-on-the-new-world/
30. Jeff Jarvis, "e-G8: A discussion about sovereignty," Buzzmachine.com, June 3, 2011, www.buzzmachine.com/2011/06/03/e-g8-a-discussion-about-sovereignty/, and "A Hippocratic oath for the internet," Buzzmachine.com, May 23, 2011, www.buzzmachine.com/2011/05/23/a-hippocratic-oath-for-the-internet/

Bibliography

Alderman, Ellen, and Caroline Kennedy. *The Right to Privacy*. New York: Vintage, 1995.

Arendt, Hannah. *The Human Condition*. Chicago: University of Chicago Press, 1958.

Bakewell, Sarah. *How to Live, or, A Life of Montaigne: In One Question and Twenty Attempts at an Answer*. New York: Other Press, 2010.

Baron, Sabrina Alcorn, Eric N. Lindquist, and Eleanor F. Shevlin, eds. *Agent of Change: Print Culture Studies After Elizabeth L. Eisenstein*. Amherst: University of Massachusetts Press, 2007.

Benkler, Yochai. *The Wealth of Networks: How Social Production Transforms Marks and Freedom*. New Haven, Conn.: Yale University Press, 2006.

boyd, danah. "Making Sense of Privacy and Publicity." SXSW. March 13, 2010.

Brandeis, Warren, and Louis D. Warren. "The Right to Privacy." Harvard Law Review 4, December 15, 1890, p. 193.

Brin, David. *The Transparent Society*. New York: Basic Books, 1998.

Calhoun, Craig, ed. *Habermas and the Public Sphere*. Cambridge, Mass.: MIT Press, 1992.

Cayle, David. *The Origins of the Modern Public* (radio series). Toronto: CBC, 2010.

Chartier, Roger, ed. *A History of Private Life: Passions of the Renaissance*, trans. Arthur Goldhammer. Cambridge, Mass.: Belknap Press, 1989.

Coupland, Douglas. *Marshall McLuhan: You Know Nothing of My Work!* New York: Atlas, 2010.

Cowan, Brian. *The Social Life of Coffee: The Emergence of the British Coffeehouse.* New Haven, Conn.: Yale University Press, 2005.

Dewar, James A. "The Information Age and the Printing Press: Looking Backward to See Ahead." Rand, 1998.

Eisenstein, Elizabeth L. *Divine Art, Infernal Machine: The Reception of Printing in the West from First Impressions to the Sense of an Ending.* Philadelphia: University of Pennsylvania Press, 2011.

———. *The Printing Press as an Agent of Change.* Cambridge, England: Cambridge University Press, 1979.

———. "An Unacknowledged Revolution Revisited." The American Historical Review 107, no. 2 (February 2002). www.historycooperative .org/journals/ahr/107.1/ah0102000087.html.

Febvre, Lucien, and Henri-Jean Martin. *The Coming of the Book: The Impact of Printing, 1450–1800,* trans. David Gerard. London: Verso, 1976.

Friedman, Lawrence. *Guarding Life's Dark Secrets: Legal and Social Controls over Reputation, Propriety, and Privacy.* Stanford, Calif.: Stanford University Press, 2007.

Gadja, Amy. "What if Samuel D. Warren Hadn't Married a Senator's Daughter? Uncovering the Press Coverage That Led to 'The Right to Privacy.'" Michigan State Law Review 35 (Spring 2008): 35–60.

Girouard, Mark. *Life in the English Country House.* New Haven, Conn.: Yale University Press, 1978.

Habermas, Jürgen. "Political Communication in Media Society." Speech before the International Communication Association. Communication Theory 16 (2006), p. 423.

———. *The Structural Transformation of the Public Sphere: An Inquiry into a Category of Bourgeois Society,* trans. Thomas Burger. Cambridge, Mass.: MIT Press, 1991.

Higgins, John H., ed. *The Raymond Williams Reader.* Malden, Mass.: Blackwell, 2001.

Hind, Dan. *The Return of the Public.* London: Verso, 2010.

Jackaway, Gwenyth L. *Media at War: Radio's Challenge to the Newspapers, 1924–1939.* Westport, Conn.: Praeger, 1995.

Johns, Adrian. "How to Acknowledge a Revolution." The American Historical Review 107, No. 1, February 2002.

———. *The Nature of the Book: Print and Knowledge in the Making.* Chicago: University of Chicago Press, 1988.

Kamvar, Sep, and Jonathan Harris, eds. *We Feel Fine: An Almanac of Human Emotion.* New York: Scribner, 2009.

Kapr, Albert. *Johann Gutenberg: The Man and his Invention,* trans. Douglas Martin. Aldershot, England: Scolar Press, 1996.

Kirkpatrick, David. *The Facebook Effect: The Inside Story of the Company That Is Connecting the World.* New York: Simon & Schuster, 2010.

Lane, Frederick S. *American Privacy: The 400-Year History of Our Most Contested Right.* Boston: Beacon Press, 2009.

Lathrop, Daniel, and Laurel Ruma. *Open Government: Collaboration, Transparency, and Participation in Practice.* Sebastopol, Calif.: O'Reilly, 2010.

Lessig, Lawrence. *Code and Other Laws of Cyberspace.* New York: Basic Books, 2006.

Levine, Rick, Christopher Locke, Doc Searls, and David Weinberger. *The Cluetrain Manifesto: The End of Business as Usual.* Cambridge, Mass.: Perseus, 2000.

Man, John. *The Gutenberg Revolution: How Printing Changed the Course of History*. London: Bantam, 2002.

Marcus, Leah. "Cyberspace Renaissance." English Literary Renaissance 25, no. 3 (September 1995), pp. 388–401.

Mayer-Schönberger, Viktor. *Delete: The Virtue of Forgetting in the Digital Age*. Princeton, N.J.: Princeton University Press, 2009.

McKeon, Michael. *The Secret History of Domesticity: Public, Private, and the Division of Knowledge*. Baltimore: Johns Hopkins, 2005.

McLuhan, Marshall, and Quentin Fiore. *The Gutenberg Galaxy*. Toronto: University of Toronto Press, 1962.

———. *The Medium Is the Massage: An Inventory of Effects*. Berkeley, Calif.: Gingko Press, 1967.

Mills, Charles Wright. *The Sociological Imagination*. Oxford, England: Oxford University Press, 1959.

Montaigne, Michel de. *Michel de Montaigne: The Complete Essays,* trans. M. A. Screech. London: Penguin, 1987.

Munson, Eve Stryker, and Catherine A. Warren, eds. *James Carey: A Critical Reader*. Minneapolis: University of Minnesota Press, 1997.

Nissenbaum, Helen. *Privacy in Context: Technology, Policy, and the Integrity of Social Life*. Stanford, Calif.: Stanford Law Books, 2010.

Noveck, Beth Simone. *Wiki Government: How Technology Can Make Government Better, Democracy Stronger, and Citizens More Powerful*. Washington, D.C.: Brookings Institution Press, 2010.

Owen, David. *Copies in Seconds: How a Lone Inventor and an Unknown Company Created the Biggest Communication Breakthrough Since Gutenberg: Chester Carlson and the Birth of the Xerox Machine*. New York: Simon & Schuster, 2004.

Pettegree, Andrew. *The Book in the Renaissance*. New Haven, Conn.: Yale University Press, 2010.

Pettitt, Tom. "Before the Gutenberg Parenthesis: Elizabethan-American Compatibilities." http://web.mit.edu/comm-forum/mit5/papers/pettitt_plenary_gutenberg.pdf.

Potter, Andrew. *The Authenticity Hoax: How We Get Lost Finding Ourselves*. New York: Harper, 2010.

Prosser, William L. "Privacy." California Law Review, 48, No. 3, August 1960.

Robinson, David G., Harlan Yu, William P. Zeller, and Edward W. Felten. "Government Data and the Invisible Hand." Yale Journal of Law & Technology, 2009.

Rosen, Jay. *The Impossible Press: American Journalism and the Decline of Public Life*. New York: New York University (dissertation), 1986.

———. *What Are Journalists For?* New Haven, Conn.: Yale University Press, 1999.

Schaar, Peter. *Das Ende der Privatsphäre: Der Web in die Überwachungsgesellschaft*. Munich, Germany: Goldman, 2007.

Sennett, Richard. *The Fall of Public Man*. New York: Norton, 1974.

Shirky, Clay. *Cognitive Surplus: Creativity and Generosity in a Connected Age*. New York: Penguin, 2010.

Sifry, Micah L. *WikiLeaks and the Age of Transparency*. New York: OR Books, 2011.

Solove, Daniel J. *The Future of Reputation: Gossip, Rumor, and Privacy on the Internet*. New Haven, Conn.: Yale University Press, 2007.

———. *Understanding Privacy*. Cambridge, Mass.: Harvard University Press, 2008.

Spacks, Patricia Ann Meyer, ed. *Privacy: Concealing the Eighteenth Century Self.* Chicago: University of Chicago Press, 2003.

Stern, Howard. *Private Parts.* New York: Pocket Books, 1993.

Tapscott, Don, and Anthony D. Williams. *Wikinomics: How Mass Collaboration Changes Everything.* New York: Portfolio, 2006.

Vaidhyanathan, Siva. "Naked in the 'Nonopticon.'" The Chronicle Review, February 15, 2008.

Veyne, Paul, ed. *A History of Private Life: From Pagan Rome to Byzantium,* trans. Arthur Goldhammer. Cambridge, Mass.: Belknap, 1987.

Warner, Michael. *Publics and Counterpublics.* New York: Zone Books, 2005.

Westin, Alan F. *Privacy and Freedom.* New York: Atheneum, 1967.

Wilson, Bronwen, and Paul Yachnin. *Making Publics in Early Modern Europe: People, Things, Forms of Knowledge.* New York: Routledge, 2010.

Index

Printed in the United States
By Bookmasters